Safety Critical Systems Handbook

A Straightforward Guide to Functional Safety: IEC 61508 (2010 Edition) and Related Standards

Including: Process IEC 61511, Machinery IEC 62061 and ISO 13849

THIRD EDITION

Previously published as: Functional Safety: A Straightforward Guide to Applying IEC 61508 and Related Standards

Dr David J Smith
Kenneth GL Simpson

AMSTERDAM • BOSTON • HEIDELBERG • LONDON • NEW YORK • OXFORD
PARIS • SAN DIEGO • SAN FRANCISCO • SINGAPORE • SYDNEY • TOKYO

Butterworth-Heinemann is an imprint of Elsevier

Butterworth-Heinemann is an imprint of Elsevier
The Boulevard, Langford Lane, Kidlington, Oxford OX5 1GB, UK
30 Corporate Drive, Suite 400, Burlington, MA 01803, USA

Editions 1 and 2 published as: Functional Safety: A Straightforward Guide to Applying IEC 61508 and Related Standards

First edition 2001
Second edition 2004
Third edition 2011

British Library Cataloguing in Publication Data
A catalogue record for this book is available from the British Library

Library of Congress Cataloging-in-Publication Data
A catalog record for this book is availabe from the Library of Congress

ISBN: 978-0-08 -096781-3

For information on all Butterworth-Heinemann publications visit our web site at books.elsevier.com

Transferred to Digital Print 2011
Printed and bound by CPI Group (UK) Ltd, Croydon, CR0 4YY

Contents

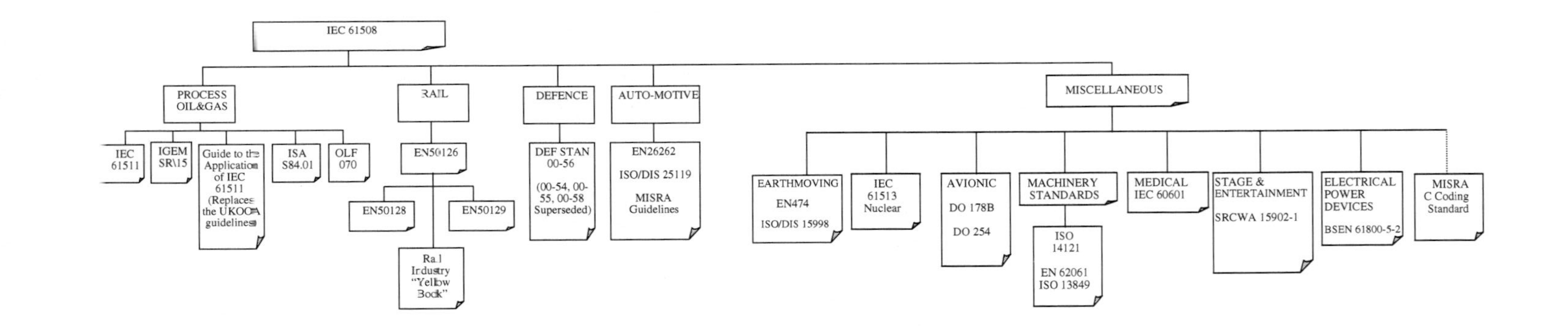
IEC 61508
PROCESS OIL&GAS
IEC 61511
IGEM SR\15
Guide to the Application of IEC 61511 (Replaces the UKOOA guidelines)
ISA S84.01
OLF 070
RAIL
EN50126
EN50128
EN50129
Ra l Irdustry "Yelbw Bock"
DEFENCE
DEF STAN 00-56
(00-54, 00-55, 00-58 Superseded)
AUTO-MOTIVE
EN26262
ISO/DIS 25119
MISRA Guidelines
MISCELLANEOUS
EARTHMOVING
EN474
ISO/DIS 15998
IEC 61513 Nuclear
AVIONIC
DO 178B
DO 254
MACHINERY STANDARDS
ISO 14121
EN 62061
ISO 13849
MEDICAL
IEC 60601
STAGE & ENTERTAINMENT
SRCWA 15902-1
ELECTRICAL POWER DEVICES
BSEN 61800-5-2
MISRA C Coding Standard

A Quick Overview

Functional safety engineering involves identifying specific hazardous failures which lead to serious consequences (e.g. death) and then establishing maximum tolerable frequency targets for each mode of failure. Equipment whose failure contributes to each of these hazards is identified and usually referred to as "safety-related". Examples are industrial process control systems, process shut down systems, rail signalling equipment, automotive controls, medical treatment equipment, etc. In other words, any equipment (with or without software) whose failure can contribute to a hazard is likely to be safety-related.

A safety-function is thus defined as a function, of a piece of equipment, that maintains it in a safe state, or brings it to a safe state, in respect of some particular hazard.

Since the publication of the first two editions of this book, in 2001 and 2004, the application of IEC 61508 has spread rapidly through most sectors of industry. Also, the process sector IEC 61511 has been published. Furthermore IEC 61508 (BS EN 16508 in the UK) has been re-issued in 2010. The opportunity has therefore been taken to update and enhance this book in the light of the authors' recent experience. There are now three chapters on industry sectors and Chapters 15 and 16 have been added to provide even more examples.

There are both **random hardware failures** which can be quantified and assessed in terms of failure rates AND **systematic failures** which cannot be quantified. Therefore it is necessary to have the concept of integrity levels so that the systematic failures can be addressed by levels of rigor in the design techniques and operating activities.

The maximum tolerable failure rate that we set, for each hazard, will lead us to an integrity target for each piece of equipment, depending upon its relative contribution to the hazard in question. These integrity targets, as well as providing a numerical target to meet, are also expressed as "safety-integrity levels" according to the severity of the numerical target. This usually involves four discrete bands of "rigor" and this is explained in Chapters 1 and 2.

SIL 4: the highest target and most onerous to achieve, requiring state of the art techniques (usually avoided)
SIL 3: less onerous than SIL 4 but still requiring the use of sophisticated design techniques

SIL 2: requiring good design and operating practice to a level such as would be found in an ISO 9001 managements system
SIL 1: the minimum level but still implying good design practice
<SIL 1: referred to (in IEC 61508 & other documents) as "not-safety related" in terms of compliance

An assessment of the design, the designer's organization and management, the operator's and the maintainer's competence and training should then be carried out in order to determine if the proposed (or existing) equipment actually meets the target SIL in question.

Overall, the steps involve:

Setting the SIL targets	Chapter 2.1
Capability to design for functional safety	Chapter 2.3
Quantitative assessment	Chapters 2, 5 and 6
Qualitative assessment	Chapters 3 and 4
Establishing competency	Chapter 2.3
As low as reasonably practicable	Chapter 2.2
Reviewing the assessment itself	Appendix 2

IEC 61508 is a generic standard which deals with the above. It can be used on its own or as a basis for developing industry sector specific standards (Chapters 8, 9 & 10). In attempting to fill the roles of being both a global template for the development of application specific standards, and being a standard in its own right, it necessarily leaves much to the discretion and interpretation of the user.

It is vital to bear in mind, however, that no amount of assessment will lead to enhanced integrity unless the assessment process is used as a tool during the design-cycle.

Now read on!

The 2010 Version of IEC 61508

The following is a brief summary of the main changes which have brought about the new, 2010, version.

Architectural Constraints (Chapter 3)

An alternative route to the "safe failure fraction" (the so called Route 1_H) requirements has been introduced (known as Route 2_H).

Route 2_H allows the "safe failure fraction" requirements to lapse providing that amount of redundancy (so called hardware fault tolerance) meets a minimum requirement AND there is adequate user based information providing failure rate data.

Security (Chapter 2)

Malevolent and unauthorized actions, as well as human error and equipment failure, can be involved in causing a hazard. They are to be taken account of, if relevant, in risk assessments.

Safety Specifications (Chapter 3)

There is more emphasis on the distinct safety requirements leading to separately defined design requirements.

Digital Communications (Chapter 3)

More detail in providing design and test requirements for "black box" and "white box" communications links.

ASICs and Integrated Circuits (Chapters 3 and 4)

More detailed techniques and measures are defined and described in Annexes to the Standard.

Safety Manual (Chapters 3 and 4)

Producers are required to provide a safety manual (applies to hardware and to software) with all the relevant safety-related information. Headings are described in Annexes to the Standard.

Synthesis of Elements (Chapter 3)

In respect of systematic failures, the ability to claim an increment of one SIL for parallel elements.

Software Properties of Techniques (Chapter 4)

New guidance on justifying the properties which proposed alternative software techniques should achieve in order to be acceptable.

Element (Appendix 8)

The introduction of a new term (similar to a subsystem).

Acknowledgements

The authors are very grateful to Mike Dodson, Independent Consultant, of Solihull, for extensive comments and suggestions and for a thorough reading of the earlier manuscript:

Many thanks to Dr Tony Foord for constructive comments on Chapters 3 and 4 and for help with the original Chapter 14.

Thanks, also, to:

Mr Paul Reeve of SIRA Certification for comments on Chapter 7.
Mr Simon Burwood, of ESC, for help with Chapter 9.
Mr Stephen Waldron, of JCB, and Mr Peter Stanton, of Railtrack, for help with Chapter 10.

PART A

The Concept of Safety Integrity

In the first chapter we will introduce the concept of functional safety and the need to express targets by means of safety integrity levels. Functional safety will be placed in context, along with risk assessment, likelihood of fatality and the cost of conformance.

The life-cycle approach, together with the basic outline of IEC 61508 (known as BS EN 61508 in the UK), will be explained.

CHAPTER 1

The Meaning and Context of Safety Integrity Targets

Chapter Outline

Safety Critical Systems Handbook. DOI: 10.1016/B978-0-08-096781-3.10001-X

1.1 Risk and the Need for Safety Targets

There is no such thing as zero risk. This is because no physical item has zero failure rate, no human being makes zero errors and no piece of software design can foresee every operational possibility.

Nevertheless public perception of risk, particularly in the aftermath of a major incident, often calls for the zero risk ideal. However, in general most people understand that this is not practicable, as can be seen from the following examples of everyday risk of death from various causes:

All causes (mid-life including medical)	1×10^{-3} pa
All accidents (per individual)	5×10^{-4} pa
Accident in the home	4×10^{-4} pa
Road traffic accident	6×10^{-5} pa
Natural disasters (per individual)	2×10^{-6} pa

Therefore the concept of defining and accepting a tolerable risk for any particular activity prevails.

The actual degree of risk considered to be tolerable will vary according to a number of factors such as the degree of control one has over the circumstances, the voluntary or involuntary nature of the risk, the number of persons at risk in any one incident and so on. This partly explains why the home remains one of the highest areas of risk to the individual in everyday life since it is there that we have control over what we choose to do and are therefore prepared to tolerate the risks involved.

A safety technology has grown up around the need to set target risk levels and to evaluate whether proposed designs meet these targets, be they process plant, transport systems, medical equipment or any other application.

In the early 1970s people in the process industries became aware that, with larger plants involving higher inventories of hazardous material, the practice of learning by mistakes (if indeed we do) was no longer acceptable. Methods were developed for identifying hazards and for quantifying the consequences of failures. They were evolved largely to assist in the decision-making process when developing or modifying plant. External pressures to identify and quantify risk were to come later.

By the mid 1970s there was already concern over the lack of formal controls for regulating those activities which could lead to incidents having a major impact on the health and safety of the general public. The Flixborough incident in June 1974, which resulted in 28 deaths, focused UK public and media attention on this area of technology. Many further events, such as that at Seveso (Italy) in 1976 through to the Piper Alpha offshore disaster and more recent

Paddington (and other) rail incidents, have kept that interest alive and have given rise to the publication of guidance and also to legislation in the UK.

The techniques for quantifying the predicted frequency of failures are just the same as those previously applied to plant availability, where the cost of equipment failure was the prime concern. The tendency in the last few years has been towards a more rigorous application of these techniques (together with third party verification) in the field of hazard assessment. They include Fault Tree Analysis, Failure Mode & Effect Analysis, Common Cause Failure Assessment and so on. These will be explained in Chapters 5 and 6.

Hazard assessment of process plant, and of other industrial activities, was common in the 1980s but formal guidance and standards were rare and somewhat fragmented. Only Section 6 of the Health and Safety at Work Act 1974 underpinned the need to do all that is reasonably practicable to ensure safety. However, following the Flixborough disaster, a series of moves (including the Seveso directive) led to the CIMAH (Control of Industrial Major Accident Hazards) regulations, 1984, and their revised COMAH form (Control of Major Accident Hazards) in 1999. The adoption of the Machinery Directive by the EU, in 1989, brought the requirement for a documented risk analysis in support of CE marking.

Nevertheless, these laws and requirements do not specify how one should go about establishing a target tolerable risk for an activity, nor do they address the methods of assessment of proposed designs nor provide requirements for specific safety-related features within design.

The need for more formal guidance has long been acknowledged. Until the mid 1980s risk assessment techniques tended to concentrate on quantifying the frequency and magnitude of consequences arising from given risks. These were sometimes compared with loosely defined target values but, being a controversial topic, such targets (usually in the form of fatality rates) were not readily owned up to or published.

EN 1050 (Principles of risk assessment), in 1996, covered the processes involved in risk assessment but gave little advice on risk reduction. For machinery control EN 954-1 (see Chapter 10) provided some guidance on how to reduce risks associated with control systems but did not specifically include PLCs (programmable logic controllers) which were separately addressed by other IEC (International Electrotechnical Commission) and CENELEC (European Committee for Standardization) documents.

The proliferation of software during the 1980s, particularly in real time control and safety systems, focused attention on the need to address systematic failures since they could not necessarily be quantified. In other words whilst hardware failure rates were seen as a credibly predictable measure of reliability, software failure rates were generally agreed not to be predictable. It became generally accepted that it was necessary to consider qualitative defenses against systematic failures as an additional, and separate, activity to the task of predicting the probability of so called random hardware failures.

In 1989, the HSE (Health and Safety Executive) published guidance which encouraged this dual approach of assuring functional safety of programmable equipment. This led to IEC work, during the 1990s, which culminated in the international safety Standard IEC 61508 – the main subject of this book. The IEC Standard is concerned with electrical, electronic and programmable safety-related systems where failure will affect people or the environment. It has a voluntary, rather than legal, status in the UK but it has to be said that to ignore it might now be seen as "not doing all that is reasonably practicable" in the sense of the Health and Safety at Work Act and a failure to show "due diligence". As use of the Standard becomes more and more widespread it can be argued that it is more and more "practicable" to use it. The Standard was revised and re-issued in 2010. Figure 1.1 shows how IEC 61508 relates to some of the current legislation.

The purpose of this book is to explain, in as concise a way as possible, the requirements of IEC 61508 and the other industry-related documents (some of which are referred to as 2nd tier guidance) which translate the requirements into specific application areas.

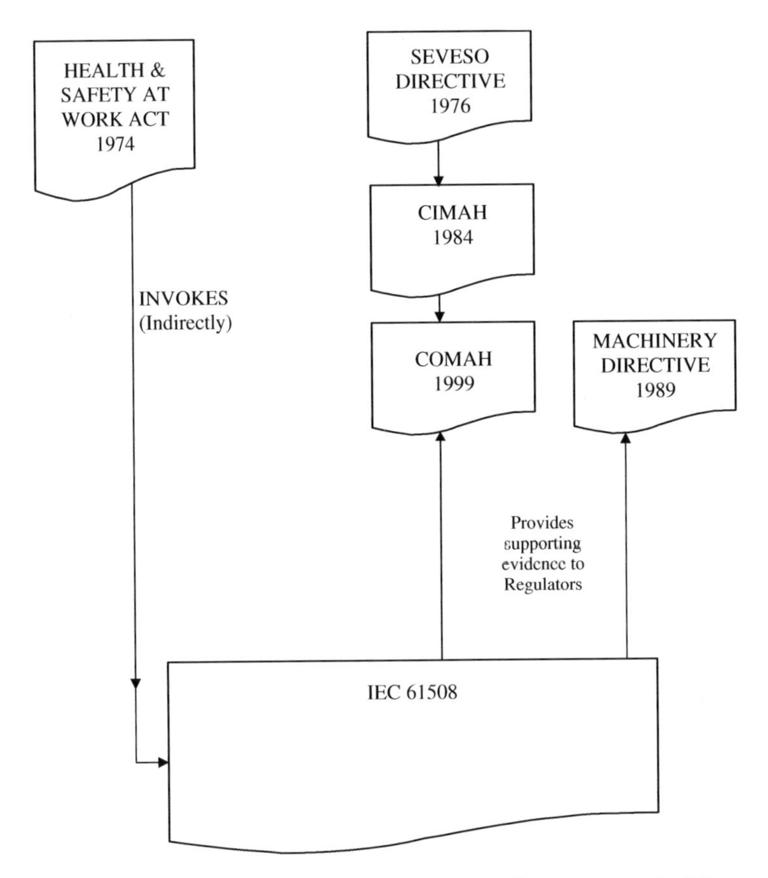

Figure 1.1: How IEC 61508 relates to some of the current legislation.

The Standard, as with most such documents, has considerable overlap, repetition, and some degree of ambiguity, which places the onus on the user to make interpretations of the guidance and, in the end, apply his/her own judgement.

The question frequently arises as to what is to be classified as safety-related equipment. The term 'safety-related' applies to any hard-wired or programmable system where a failure, singly or in combination with other failures/errors, could lead to death, injury or environmental damage. The terms "safety-related" and "safety-critical" are often used and the distinction has become blurred. "Safety-critical" has tended to be used where failure alone, of the equipment in question, leads to a fatality or increase in risk to exposed people. "Safety-related" has a wider context in that it includes equipment in which a single failure is not necessarily critical whereas coincident failure of some other item leads to the hazardous consequences.

A piece of equipment, or software, cannot be excluded from this safety-related category merely by identifying that there are alternative means of protection. This would be to pre-judge the issue and a formal safety integrity assessment would still be required to determine whether the overall degree of protection is adequate.

1.2 Quantitative and Qualitative Safety Targets

In an earlier paragraph we introduced the idea of needing to address safety-integrity targets both quantitatively and qualitatively:

> **Quantitatively:** where we predict the frequency of hardware failures and compare them with some tolerable risk target. If the target is not satisfied then the design is adapted (e.g. provision of more redundancy) until the target is met.
> **Qualitatively:** where we attempt to minimize the occurrence of systematic failures (e.g. software errors) by applying a variety of defenses and design disciplines appropriate to the severity of the tolerable risk target.

It is important to understand why this twofold approach is needed. Prior to the 1980s, system failures could usually be identified as specific component failures (e.g. relay open circuit, capacitor short circuit, motor fails to start). However, since then the growth of complexity (including software) has led to system failures of a more subtle nature whose cause may not be attributable to a catastrophic component failure. Hence we talk of:

> **Random hardware failures:** which are attributable to specific component failures and to which we attribute failure rates. The concept of "repeatability" allows us to model proposed systems by means of associating past failure rates of like components together to predict the performance of the design in question.
> and
> **Systematic failures:** which are not attributable to specific component failures and are therefore unique to a given system and its environment. They include design tolerance/

timing related problems, failures due to inadequately assessed modifications and, of course, software. Failure rates cannot be ascribed to these incidents since they do not enable us to predict the performance of future designs.

Quantified targets can therefore be set for the former (random hardware failures) but not for the latter. Hence the concept emerges of an arbitrary number of levels of rigor/excellence in the control of the design and operations. The ISO 9001 concept of a qualitative set of controls is somewhat similar and is a form of single "SIL". In the Functional Safety profession the practice has been to establish four such levels of rigor according to the severity of the original risk target.

During the 1990s this concept of safety-integrity levels (known as SILs) evolved and is used in the majority of documents in this area. The concept is to divide the "spectrum" of integrity into four discrete levels and then to lay down requirements for each level. Clearly, the higher the SIL then the more stringent become the requirements. In IEC 61508 (and in most other documents) the four levels are defined as shown in Table 1.1.

Note that had the high demand SIL bands been expressed as "per annum" then the tables would appear numerically similar. However, being different parameters, they are NOT even the same dimensionally. Thus the "per hour" units are used to minimize confusion.

The reason for there being effectively two tables (high and low demand) is that there are two ways in which the integrity target may need to be described. The difference can best be understood by way of examples.

Consider the motor car brakes. It is the rate of failure which is of concern because there is a high probability of suffering the hazard immediately each failure occurs. Hence we have the middle column of Table 1.1.

On the other hand, consider the motor car air bag. This is a low demand protection system in the sense that demands on it are infrequent (years or tens of years apart). Failure rate alone is of little use to describe its integrity since the hazard is not incurred immediately each failure occurs and we therefore have to take into consideration the test interval. In other words, since the demand is infrequent, failures may well be dormant and persist during the test interval.

Table 1.1: Safety Integrity Levels (SILs).

Safety integrity level	High demand rate (dangerous failures/hr)	Low demand rate (probability of failure on demand)
4	$>= 10^{-9}$ to $< 10^{-8}$	$>= 10^{-5}$ to $< 10^{-4}$
3	$>= 10^{-8}$ to $< 10^{-7}$	$>= 10^{-4}$ to $< 10^{-3}$
2	$>= 10^{-7}$ to $< 10^{-6}$	$>= 10^{-3}$ to $< 10^{-2}$
1	$>= 10^{-6}$ to $< 10^{-5}$	$>= 10^{-2}$ to $< 10^{-1}$

What is of interest is the combination of failure rate and down time and we therefore specify the probability of failure on demand (PFD): hence the right hand column of Table 1.1.

In IEC 61508 (clause 3.5.14 of part 4) the high demand definition is called for when the demand on a safety related function is greater than once per annum and the low demand definition when it is less frequent.

In Chapter 2 we will explain the ways of establishing a target SIL and it will be seen that the IEC 61508 Standard then goes on to tackle the two areas of meeting the quantifiable target and addressing the qualitative requirements separately.

A frequent misunderstanding is to assume that if the qualitative requirements of a particular SIL are observed the numerical failure targets, given in Table 1.1, will automatically be achieved. This is most certainly not the case since the two issues are quite separate. The quantitative targets refer to random hardware failures and are dealt with in Chapters 5 and 6. The qualitative requirements refer to quite different types of failure whose frequency is NOT quantified and are thus dealt with separately. The assumption, coarse as it is, is that by spreading the rigor of requirements across the range SIL 1 − SIL 4, which in turn covers the credible range of achievable integrity, the achieved integrity is likely to coincide with the measures applied.

A question sometimes asked is: If the quantitative target is met by the predicted random hardware failure probability then what allocation should there be for the systematic (software) failures? The target is to be applied equally to random hardware failures and to systematic failures. In other words the numerical target is not divided between the two but applied to the random hardware failures. The corresponding SIL requirements are then applied to the systematic failures. In any case, having regard to the accuracy of quantitative predictions (see Chapter 6), the point may not be that important. The 2010 version implies this in 7.4.5.1 of Part 2.

The following should be kept in mind:

SIL 1: is relatively easy to achieve especially if ISO 9001 practices apply throughout the design providing that Functional Safety Capability is demonstrated.
SIL 2: is not dramatically harder than SIL 1 to achieve although clearly involving more review and test and hence more cost. Again, if ISO 9001 practices apply throughout the design, it should not be difficult to achieve.

(SILs 1 and 2 are not dramatically different in terms of the life-cycle activities)

SIL 3: involves a significantly more substantial increment of effort and competence than is the case from SIL 1 to SIL 2. Specific examples are the need to revalidate the system following design changes and the increased need for training of operators. Cost and time will be a significant factor and the choice of vendors will be more limited by lack of ability to provide SIL 3 designs.

SIL 4: involves state of the art practices including "formal methods" in design. Cost will be extremely high and competence in all the techniques required is not easy to find. There is a considerable body of opinion that SIL 4 should be avoided and that additional levels of protection should be preferred.

It is reasonable to say that the main difference between the SILs is the quantification of random hardware failures and the application of the Safe Failure Fraction rules (see Chapter 3). The qualitative requirements for SILs 1 and 2 are very similar, as are those for SILs 3 and 4. The major difference is in the increment of rigor between SIL 2 and SIL 3.

Note, also, that as one moves up the SILs the statistical implications of verification become more onerous whereas the assessment becomes more subjective due to the limitations of the data available for the demonstration.

1.3 The Life-cycle Approach

Section 7.1 of Part 1

The various life-cycle activities and defenses against systematic failures, necessary to achieve functional safety, occur at different stages in the design and operating life of an equipment. Therefore it is considered a good idea to define (that is to say describe) a life-cycle.

IEC 61508 is based on a safety life-cycle approach, describes such a model, and identifies activities and requirements based on it. It is important to understand this because a very large proportion of safety assessment work has been (and often still is) confined to assessing whether the proposed design configuration (architecture) meets the target failure probabilities (dealt with later in Chapters 5 and 6 of this book). Because of systematic failures, modern guidance (especially IEC 61508) requires a much wider approach involving control over all of the life-cycle activities that influence safety-integrity.

Figure 1.2 shows a simple life-cycle very similar to the one shown in the Standard. It has been simplified for the purposes of this book.

As far as IEC 61508 is concerned this life-cycle applies to all electrical and programmable aspects of the safety-related equipment. Therefore if a safety-related system contains an E/PE element then the Standard applies to all the elements of system, including mechanical and pneumatic equipment. There is no reason, however, why it should not also be used in respect of "other technologies" where they are used to provide risk reduction. For that reason the Gas Industry document IGEM/SR/15 is entitled "Integrity of safety-related systems in the gas industry" in order to include all technologies.

The IEC 61508 headings are summarized in the following pages and also map to the headings in Chapters 3 and 4. This is because the Standard repeats the process for systems hardware (Part 2) and for software (Part 3). IEC 65108 Part 1 lists these in its "Table 1" with associated

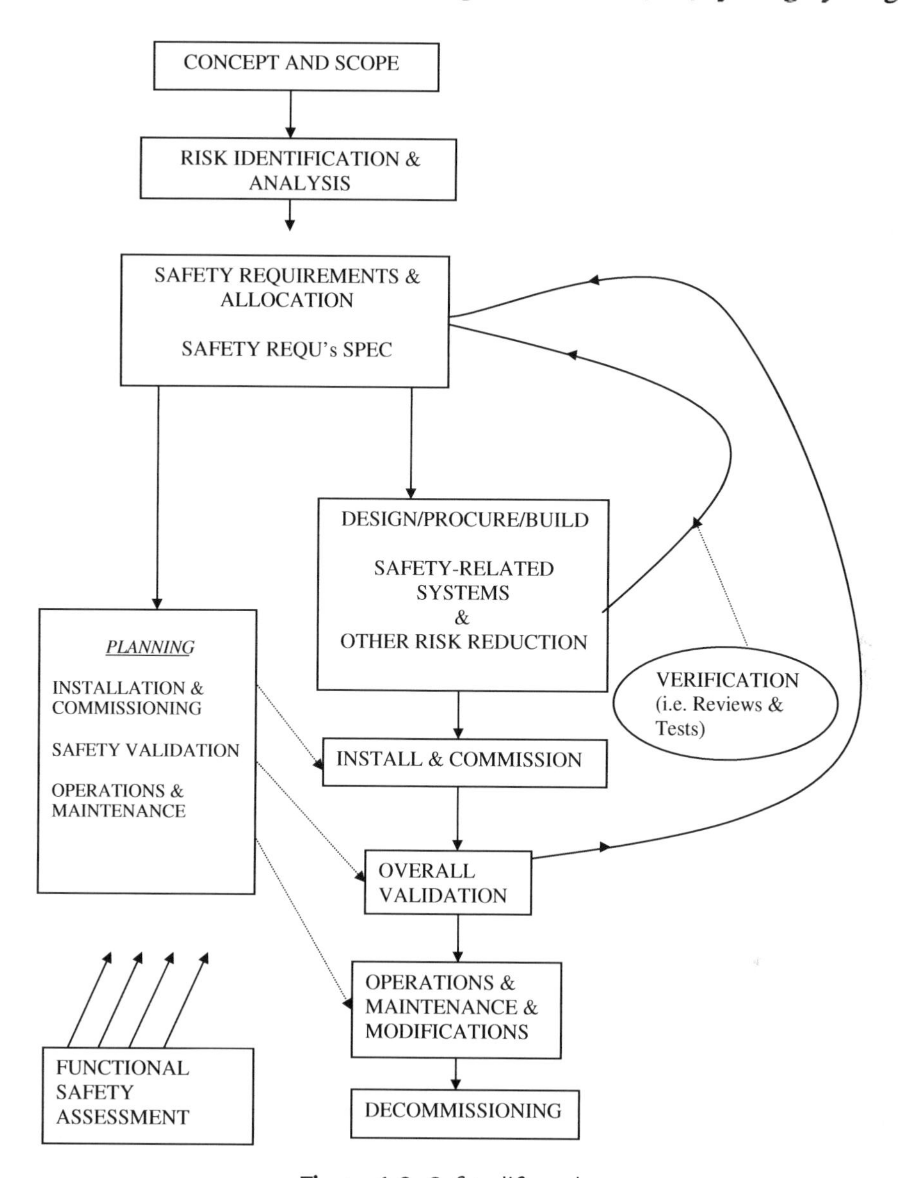

Figure 1.2: Safety life-cycle.

paragraphs of text. The following text refers to the items in IEC 61508 Part 1 Table 1 and provides the associated paragraph numbers.

Concept and scope [Part 1 – 7.2 and 7.3]

Defines exactly what is the EUC (equipment under control) and the part(s) being controlled. Understands the EUC boundary and its safety requirements. Scopes the extent of the hazard and identification techniques (e.g. HAZOP). Requires a safety plan for all the life-cycle activities.

Hazard and risk analysis [Part 1 — 7.4]

This involves the quantified risk assessment by considering the consequences of failure (often referred to as HAZAN).

Safety requirements and allocation [Part 1 — 7.5 and 7.6]

Here we address the WHOLE SYSTEM and set maximum tolerable risk targets and allocate failure rate targets to the various failure modes across the system. Effectively this defines what the safety function is by establishing what failures are protected against and how. Thus the safety functions are defined and EACH has its own SIL (see Chapter 2).

Plan operations and maintenance [Part 1 — 7.7]

What happens in operations, and during maintenance, can effect functional safety and therefore this has to be planned. The effect of human error is important here as will be covered in Chapter 5. This also involves recording actual safety-related demands on systems as well as failures.

Plan the validation [Part 1 — 7.8]

Here we plan the overall validation of all the functions. It involves pulling together the evidence from the all the verifications (i.e. review and test) activities into a coherent demonstration of conformance to the safety-related requirements.

Plan installation and commissioning [Part 1 — 7.9]

What happens through installation and commissioning can effect functional safety and therefore this has to be planned. The effect of human error is important here as will be shown in Chapter 5.

The safety requirements specification [Part 1 — 7.10]

Describes all the safety functions in detail.

Design and build the system [Part 1 — 7.11 and 7.12]

This is called "realization" in IEC 61508. It means creating the actual safety systems be they electrical, electronic, pneumatic, and/or other failure protection levels (e.g. physical bunds or barriers).

Install and commission [Part 1 — 7.13]

Implement the installation and create records of events during installation and commissioning, especially failures.

Validate that the safety-systems meet the requirements [Part 1 — 7.14]

This involves checking that all the allocated targets (above) have been met. This will involve a mixture of predictions, reviews and test results. There will have been a validation plan (see above) and there will need to be records that all the tests have been carried out and recorded for both hardware and software to see that they meet the requirements of the target SIL. It is important that the system is re-validated from time to time during its life, based on recorded data.

Operate, maintain, and repair [Part 1 — 7.15]

Clearly operations and maintenance (already planned above) are important. Documentation, particularly of failures, is important.

Control modifications [Part 1 — 7.16]

It is also important not to forget that modifications are, in effect, re-design and that the life-cycle activities should be activated as appropriate when changes are made.

Disposal [Part 1 — 7.17]

Finally, decommissioning carries its own safety hazards which should be taken into account.

Verification [Part 1 — 7.18]

Demonstrating that all life-cycle stage deliverables were met in use.

Functional safety assessments [Part 1 — 8]

Carry out assessments to demonstrate compliance with the target SILs (see Chapter 2.3 of this book for the extent of independence according to consequences and SIL).

1.4 Steps in the Assessment Process

The following steps are part of the safety life-cycle (functional safety assessment).

Step 1. Establish Functional Safety Capability (i.e. Management)

Whereas Steps 2—7 refer to the assessment of a system or product, there is the requirement to establish the FUNCTIONAL SAFETY CAPABILITY of the assessor and/or the design organization. This is dealt with in Chapter 2.3 and by means of Appendix 1.

Step 2. Establish a Risk Target

ESTABLISH THE RISK TO BE ADDRESSED by means of techniques such as formal hazard identification or HAZOP whereby failures and deviations within a process (or equipment) are studied to assess outcomes. From this process one or more hazardous events may be revealed which will lead to death or serous injury.

SET MAXIMUM TOLERABLE FAILURE RATES by carrying out a quantified risk assessment based on a maximum tolerable probability of death or injury, arising from the event in question. This is dealt with in the next Chapter and takes into account how many simultaneous risks to which one is exposed in the same place, the number of fatalities and so on.

Step 3. Identify the Safety Related Function(s)

For each hazardous event it is necessary to understand what failure modes will lead to it. In this way the various elements of protection (e.g. control valve AND relief valve AND slamshut valve) can be identified. The safety protection system for which a SIL is needed can then be identified.

Step 4. Establish SILs for the Safety-related Elements

Both the NUMERICAL ASSESSMENT, LOPA and RISK GRAPH methods are described in Chapter 2 and examples are given in Chapter 13.

Step 5. Quantitative Assessment of the Safety-related System

Reliability modeling is needed to assess the failure rate or probability of failure on demand of the safety-related element or elements in question. This can then be compared with the target set in Step 3. Chapters 5 and 6 cover the main techniques.

Step 6. Qualitative Assessment Against the Target SILs

The various requirements for limiting systematic failures are more onerous as the SIL increases. These cover many of the life-cycle activities and are covered in Chapters 3 and 4.

Step 7. Establish ALARP

It is not sufficient to establish, in Step 4, that the quantitative failure rate (or the PFD) has been met. Design improvements which reduce the failure rate (until the Broadly Acceptable failure rate is met) should be considered and an assessment made as to whether these are "as low as reasonably practicable". This is covered in Chapter 2.2.

It is worth noting, at this point, that conformance to a SIL requires that all the Steps are met. If the quantitative assessment (Step 5) indicates a given SIL then this can only be claimed if the qualitative requirements (Step 6) are also met.

Part 1 clause 8 of IEC 61508 (Functional Safety Assessment) addresses this area. FSA should be done at all lifecycle phases (not just Phase 9, Realization). There are minimum levels of independence of the assessment team from the system/company being assessed, depending on the SIL involved. In summary these are:

SIL	Consequence	Assessed by
4	Minor injury	Independent organization
3*	Severe injury or one death	Independent department
2*	More than one death**	Independent person
1	Many deaths **	Independent person

*Add one level if there is lack of experience, unusual complexity or novel design.
**Not quantified in the standard.

Typical headings in an assessment report would be:

- Hazard scenarios and associated failure modes
- SIL targeting
- Random hardware failures
- ALARP
- Architectures (SFF)
- Life-cycle activities
- Functional safety capability
- Recommendations.

1.5 Costs

The following questions are often asked:

> " What is the cost of applying IEC 61508?"
> " What are the potential savings arising from its use?"
> " What are the potential penalty costs of ignoring it?"

1.5.1 Costs of Applying the Standard

Although costs will vary considerably, according to the scale and complexity of the system or project, the following typical resources have been expended in meeting various aspects of IEC 61508.

> Full Functional Safety Capability (now called Functional Safety Management) including implementation on a project or product – 30 to 60 man-days + several £'000 for certification by an accredited body (i.e. SIRA).
> Product or Project Conformance (to the level of third-party independent assessment) – 10 to 20 man-days + a few £'000 consultancy.

Elements within this can be identified as follows:

> Typical SIL targeting with random hardware failures assessment and ALARP – 2 to 6 man-days.

Assessing the safe failure fraction of an instrument (one or two failure modes) – 1 to 3 man-days.
Bringing an ISO 9001 management system up to IEC61508 functional safety capability – 5 man-days for the purpose of a product demonstration where evidence of only random hardware failures and safe failure fraction are being offered, 20 to 50 man-days for the purpose of an accredited Functional Safety Capability certificate.

1.5.2 Savings From Implementing the Standard

For some time there has an intangible but definite benefit due to enhanced credibility in the market place. Additional sales vis à vis those who have not demonstrated conformance are likely. However, the majority of instrument and system providers now see it as necessary to demonstrate conformance to some SIL and thus it becomes a positive disadvantage not to do so.

Major savings are purported to be in reduced maintenance for those (often the majority) systems which are given low SIL targets. This also has the effect of focusing the effort on the systems with higher SIL targets.

1.5.3 Penalty Costs From Not Implementing the Standard

The manufacturer and the user will be involved in far higher costs of retrospective redesign if subsequent changes are needed to meet the maximum tolerable risk.

The user could face enormous legal costs in the event of a major incident which invokes the H&SW Act especially if the Standard had not been applied when it was reasonably practicable to have done so.

1.6 The Seven Parts of IEC 61508

Now that we have introduced the concept of safety integrity levels and described the life-cycle approach it is now appropriate to describe the structure of the IEC 61508 Standard. Parts 1–3 are the main parts (Figure 1.3) and parts 4–7 provide supplementary material.

The general strategy is to establish SIL targets, from hazard and risk analysis activities, and then to design the safety-related equipment to an appropriate integrity level taking into account random and systematic failures and also human error.

Examples of safety-related equipment might include:

Shutdown systems for processes
Interlocks for dangerous machinery
Fire and gas detection

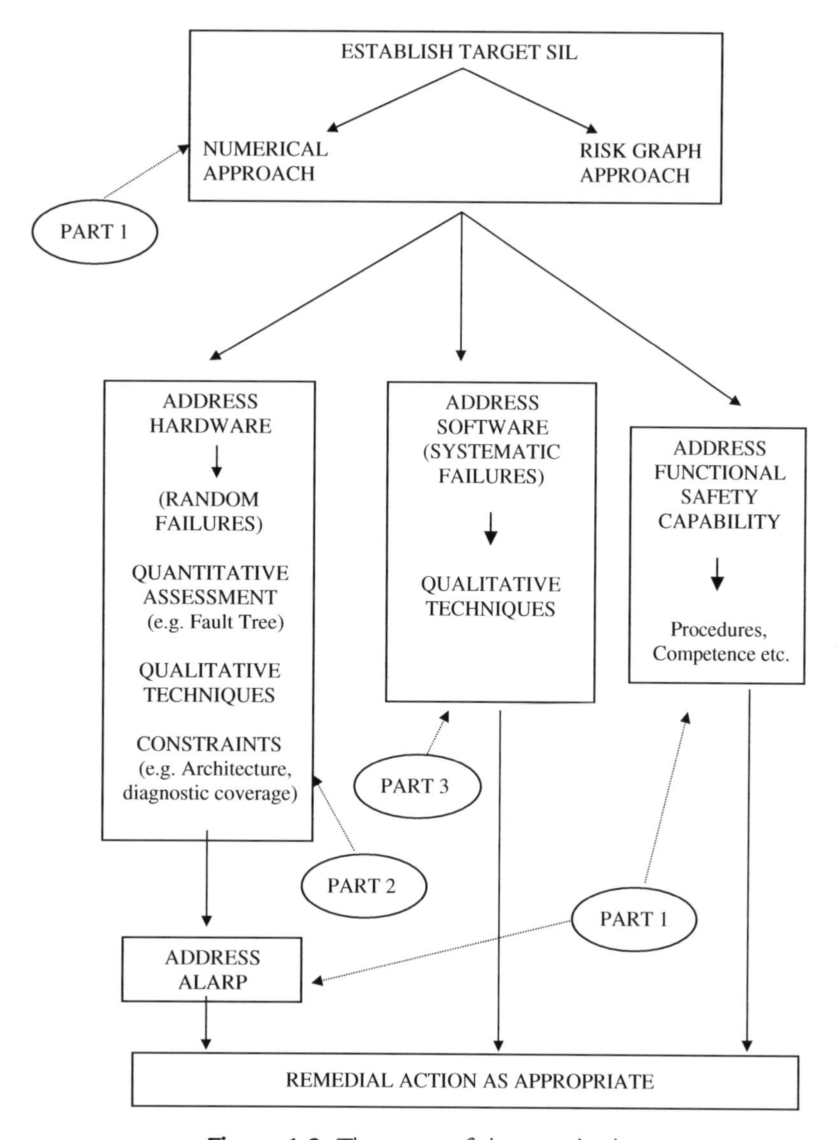

Figure 1.3: The parts of the standard.

Instrumentation
Programmable controllers
Railway signaling
Boiler and burner controls
Industrial machinery
Avionic systems
Leisure items (e.g. fairground rides)
Medical equipment (e.g. oncology systems).

Part 1 is called "General Requirements". In actual fact it covers:

(i) General functional safety management, dealt with in Chapter 2 and Appendix 1 of this book. This is the management system (possibly described in one's quality management system) which lays down the activities, procedures and skills necessary to carry out the business of risk assessment and of designing to meet integrity levels.
(ii) The life-cycle, explained above, and the requirements at each stage, are central to the theme of achieving functional safety. It will dominate the structure of several of the following Chapters and Appendices.
(iii) The definition of SILs and the need for a hazard analysis in order to define a SIL target.
(iv) The need for competency criteria for people engaged in safety-related work, also dealt with in Chapter 2 of this book.
(v) Levels of independence of those carrying out the assessment. The higher the SIL the more independent should be the assessment.
(vi) There is an Annex in Part 1 (informative only) providing a sample document structure for a safety-related design project.

Part 2 is called "Requirements for E/E/PES safety-related systems". What this actually means is that Part 2 is concerned with the hardware, rather than the software, aspects of the safety-related system. It covers:

(i) The life-cycle activities associated with the design and realization of the equipment including defining safety requirements, planning the design, validation, verification, observing architectural constraints, fault tolerance, test, subsequent modification (which will be dealt with in Chapter 3).
(ii) The need to assess (i.e. predict) the quantitative reliability (vis à vis random hardware failures) against the SIL targets in Table 1.1. This is the reliability prediction part of the process and is covered in Chapters 5 and 6.
(iii) The techniques and procedures for defending against systematic hardware failures.
(iv) Architectural constraints vis à vis the amount of redundancy applicable to each SIL. Hence, even if the above reliability prediction indicates that the SIL is met, there will still be minimum levels of redundancy. This could be argued as being because the reliability prediction will only have addressed random hardware failures (in other words those present in the failure rate data) and there is still the need for minimum defenses to tackle the systematic failures.
(v) Some of the material is in the form of annexes.

Chapter 3 of this book is devoted to summarizing Part 2 of IEC 61508.

Part 3 is called "Software requirements". As the title suggests this addresses the activities and design techniques called for in the design of the software. It is therefore about systematic failures and no quantitative prediction is involved.

(i) Tables indicate the applicability and need for various techniques at each of the SILs.
(ii) Some of the material is in the form of annexes.

Chapter 4 of this book is devoted to summarizing Part 3 of IEC 61508.

Part 4 is called "Definitions and abbreviations". This book does not propose to offer yet another list of terms and abbreviations beyond the few terms in Appendix 8. In this book the terms are hopefully made clear as they are introduced.

Part 5 is called "Examples of methods for the determination of safety-integrity levels".

As mentioned above, the majority of Part 5 is in the form of seven Annexes which are informative rather than normative:

(i) Annex A covers the general concept of the need for risk reduction through to the allocation of safety requirements, which is covered in Chapter 2 of this book.
(ii) Annex B covers methods for determining safety integrity level targets.
(iii) Annex C covers the application of the ALARP (as low as reasonably practicable) principle, which is covered in Chapter 2.2 of this book.
(iv) Annex D covers the mechanics of quantitatively determining the SIL levels, which is covered in Chapter 2.1 of this book.
(v) Annex E covers a qualitative method (risk graph) of establishing the SIL levels, which is also covered in Chapter 2 of this book.
(vi) Annex F covers Semi-quantitative LOPA (chapter 2 of this book).
(vii) Annex G describes an alternative qualitative method, "Hazardous event severity matrix".

Part 6 is called "Guidelines on the application of Part 2 and Part 3". This consists largely of informative annexes which provide material on:

(i) Calculating hardware failure probabilities (low and high demand).
(ii) Common cause failure, which is covered in Chapter 5 of this book.
(iii) Diagnostic coverage, which is covered in Chapter 3 of this book.
(iv) Applying the software requirements tables (of Part 3) for SILs 2 and 3, which is covered in Chapter 4 of this book.

As mentioned above, the majority of Part 6 is in the form of Annexes which are informative rather than normative.

Part 7 is called "Overview of techniques and measures". This is a reference guide to techniques and measures and is cross-referenced from other parts of the Standard. This book does not repeat that list but attempts to explain the essentials as it goes along.

The basic requirements are summarized in Figure 1.4.

TARGETING INTEGRITY (SILs)
ASSESSING RANDOM HARDWARE FAILURES
MEETING ALARP
ASSESSING ARCHITECTURES
MEETING THE LIFE-CYCLE REQUIREMENTS
HAVING THE FUNCTIONAL CAPABILITY TO ACHIEVE THE ABOVE

Figure 1.4: Summary of the requirements.

CHAPTER 2

Meeting IEC 61508 Part 1

Chapter Outline

Safety Critical Systems Handbook. DOI: 10.1016/B978-0-08-096781-3.10002-1

Part 1 of the Standard addresses the need for:

- Setting Integrity (SIL) targets
- The ALARP concept (by inference)
- Capability to design, operate and maintain for functional safety
- Establishing competency
- Hierarchy of documents.

The following sections summarize the main requirements:

2.1 Establishing Integrity Targets

Assessing quantified integrity targets is an essential part of the design process (including retrospective safety studies). This leads to:

- A quantified target against which one predicts the rate of random hardware failures and establishes ALARP
- A SIL band for mandating the appropriate rigor of life cycle activities.

The following paragraphs describe how a SIL target is established.

2.1.1 The Quantitative Approach

(a) Maximum tolerable risk

In order to set a quantified safety integrity target, a target Maximum Tolerable Risk is needed. It is therefore useful to be aware of the following rates:

All accidents (per individual)	5×10^{-4} pa
Natural disasters (per individual)	2×10^{-6} pa
Accident in the home	4×10^{-4} pa
Worst case maximum tolerable risk in HSE R2P2 document	10^{-3} pa
"Very low risk" as described in HSE R2P2 document (i.e. boundary between Tolerable and Broadly Acceptable)	10^{-6}pa

"Individual risk" is the frequency of fatality for a hypothetical person in respect of a specific hazard. This is different from "societal risk", which takes account of multiple fatalities. Society has a greater aversion to multiple fatalities than single ones in that killing 10 people in a single incident is perceived as worse than 10 separate single fatalities.

Table 2.1 shows the limits of tolerability for "individual risk" and is based on a review of HSE's "Reducing risk, protecting people, 2001 (R2P2)" and HSG87. The former indicates a maximum tolerable risk to an employee of 10^{-3} per annum for all risks combined. The actual risk of accidents at work per annum is well below this. Generally, guidance documents

Table 2.1: Target individual risks.

	HSE R2P2	Generally used for functional safety
Maximum Tolerable Individual Risk (per annum)		
Employee	10^{-3}	10^{-4}
Public	10^{-4}	10^{-5}
Broadly Acceptable Risk (per annum)		
Employee and public	10^{-6}	10^{-6}

recommend a target of 10^{-4} per annum for all process related risks combined, leaving a margin to allow for other types of risk.

At the lower end of the risk scale, a Broadly Acceptable Risk is nearly always defined. This is the risk below which one would not, normally, seek further risk reduction. It is approximately 2 orders of magnitude less than the total of random risks to which one is exposed in everyday life.

There is a body of opinion that multiple fatalities should also affect the choice of Maximum Tolerable Individual Risk. The targets in Table 2.2 reflect an attempt to take account of societal risk concerns in a relatively simple way by adjusting the Individual Risk targets from Table 2.1. More complex calculations for societal risk (involving F—N curves) are sometimes addressed by specialists as are adjustments for particularly vulnerable sections of the community (disabled, children etc). These are not addressed in this book.

The location, i.e. site or part of a site, for which a risk is being addressed may be exposed to multiple potential sources of risk. The question arises as to how many potential separate hazards an individual (or group) in any one place and time is exposed to. Therefore, in the event of exposure to several hazards at one time, one should seek to allow for this by specifying a more stringent target for each hazard. For example, a study addressing a multi-risk installation might need to take account of an order of magnitude of sources of risk. On the other hand, an assessment of a simple district pressure regulator valve for the local distribution of natural gas implies a limited number of sources of risk (perhaps only one).

A typical assessment confined to employees on a site might use the recommended 10^{-4} pa maximum tolerable risk (for 1—2 fatalities) but may address 10 sources of risk to an

Table 2.2: Target multiple fatality risks.

	1—2 fatalities	3—5 fatalities	6 or more fatalities
Maximum Tolerable Individual Risk (per annum)			
Employee	10^{-4}	3×10^{-5}	10^{-5}
Public	10^{-5}	3×10^{-6}	10^{-6}
Broadly Acceptable Risk (per annum)			
Employee and public	10^{-6}	3×10^{-7}	10^{-7}

individual in a particular place. Thus, an average of 10^{-5} pa would be used as the Maximum Tolerable Risk across the 10 hazards and, therefore, for each of the 10 safety functions involved. By the same token, the Broadly Acceptable Risk would be factored from 10^{-6} pa to 10^{-7} pa.

The question arises of how long an individual is exposed to a risk. Earlier practice has been to factor the maximum tolerable failure rate by the proportion of time it offers the risk (for example, an enclosure which is only visited 2 hours per week). However, that approach would only be valid if persons (on-site) suffered no other risk outside that 2 hours of his/her week. Off-site the argument might be different in that persons may well only be at risk for a proportion of the time. Thus, for on-site personnel, the proportion of employee exposure time should be taken as the total working proportion of the week.

Table 2.3 caters for the lesser consequence of injury. Targets are set in the same manner and integrity assessment is carried out as for fatality. In general, an order of magnitude larger rates are used for the targets.

In any event, the final choice of Maximum Tolerable Risk (in any scenario) forms part of the "safety argument" put forward by a system user. There are no absolute rules but the foregoing provides an overview of current practice.

Table 2.3: Target individual risks for injury.

Maximum Tolerable Risk (per annum)	
Employee	10^{-3}
Public	10^{-4}
Broadly Acceptable Risk (per annum)	
Employee and public	10^{-5}

(b) Maximum tolerable failure rate

This involves factoring the Maximum Tolerable Risk according to totally external levels of protection and to factors which limit the propagation to fatality of the event. Table 2.4 gives examples of the elements which might be considered. These are not necessarily limited to the items described below and the analyst(s) must be open ended in identifying and assessing the factors involved.

The maximum tolerable failure rate is then targeted by taking the maximum tolerable risk and factoring it according to the items assessed. Thus, for the examples given in Table 2.4 (assuming a 10^{-5} pa involuntary risk):

$$\text{Maximum Tolerable Failure Rate} = 10^{-5}\ \text{pa}/(0.6 \times 0.2 \times 0.7 \times 0.25 \times 0.9 \times 0.25)$$
$$= 2.1 \times 10^{-3}\ \text{pa}$$

Table 2.4: Factors leading to the Maximum Tolerable Failure Rate.

Factor involving the propagation of the incident or describing an independent level of protection	Probability (example)	This column is used to record arguments, justifications, references etc. to support the probability used
The profile of time at risk	60%	Quantifying whether the scenario can develop. This may be <100% as for example if: • flow, temp, pressure etc profiles are only sufficient at specific times, for the risk to apply • the process is only in use for specific periods.
Unavailability of separate mitigation fails (i.e. another level of protection)	20%	Mitigation outside the scope of this study and not included in the subsequent modeling which assesses whether the system meets the risk target. Examples are: • α down stream temp, pressure etc. measurement leading to manual intervention • a physical item of protection (for example, vessel; bund) not included in the study.
Probability of the scenario developing	70%	Examples are: • the vessel/line will succumb to the over-temp, over pressure etc. • the release has an impact on the passing vehicle.
Person(s) exposed (i.e. being at risk)	25%	Proportion of time during which some person or persons are close enough to be at risk should the event propagate. Since a person may be exposed to a range of risks during the working week, this factor should not be erroneously reduced to the proportion of time exposed to the risk in question. If that were repeated across the spectrum of risks then each would be assigned an artificially optimistic target. The working week is approximately 25% of the time and thus that is the factor which would be anticipated for an on-site risk. In the same way, an off-site risk may only apply to a given individual for a short time.
Probability of subsequent ignition	90%	Quantifying whether the released material ignites/explodes.
Fatality ensues	25%	The likelihood that the event, having developed, actually leads to fatality.

Example

A gas release (e.g. a natural gas holder over-fill) is judged to be a scenario leading to a single on-site fatality and three offsite fatalities. Both on and off site, person(s) are believed to be exposed to that one risk from the installation.

On site

Proportion of time system can offer the risk	75%	40 weeks pa
Probability of ignition	5%	Judgement
Person at risk	25%	Working week i.e. 42 hrs / 168 hrs
Probability of fatality	75%	Judgement

From Table 2.2, the maximum tolerable risk is 10^{-4} pa. Thus, the maximum tolerable failure rate (leading to the event) is calculated as:

$$10^{-4}\ \text{pa}/\ (0.75 \times 0.05 \times 0.25 \times 0.75) = 1.4 \times 10^{-2}\text{pa}$$

Off site

Proportion of time system can offer the risk	75%	40 weeks pa
Probability of ignition	5%	Judgement
Person(s) at risk	33%	Commercial premises adjoin
Probability of 3 fatalities	10%	Offices well protected by embankments

From Table 2.2 the maximum tolerable risk is 3×10^{-6} pa. Thus the maximum tolerable failure rate (leading to the event) is calculated as:

$$3 \times 10^{-6}\text{pa}/(0.75 \times 0.05 \times 0.33 \times 0.1) = 2.4 \times 10^{-3}\text{pa}$$

Thus, 2.4×10^{-3} pa, being the more stringent of the two, is taken as the maximum tolerable failure rate target.

(c) Safety integrity levels (SILs)

Notice that only now is the SIL concept introduced. The foregoing is about risk targeting but the practice of jumping immediately to a SIL target is a dangerous approach.

Furthermore, it is necessary to understand why there is any need for a SIL concept when we have numerical risk targets against which to assess the design. If the assessment were to involve only traditional reliability prediction, wherein the predicted hardware reliability is compared with a target, there would be no need for the concept of discrete SILs. However, because the rigor of adherence to design/quality assurance activities cannot be quantified, a number of discrete levels of "rigor", which cover the credible range of integrity, are described. The practice is to divide the spectrum of integrity targets into four levels (see Chapter 1).

Consider the following examples:

Simple example (low demand)

As a simple example of selecting an appropriate SIL, assume that the maximum tolerable frequency for an involuntary risk scenario (e.g. customer killed by explosion) is 10^{-5} pa (A) (see Table 2.1). Assume that 10^{-2} (B) of the hazardous events in question lead to fatality. Thus the maximum tolerable failure rate for the hazardous event will be $C = A/B = 10^{-3}$ pa. Assume that a fault tree analysis predicts that the unprotected process is only likely to achieve a failure rate of 2×10^{-1} pa (D) (i.e. 1/5 years). The FAILURE ON DEMAND of the safety system would need to be $E = C/D = 10^{-3}/2 \times 10^{-1} = 5 \times 10^{-3}$. Consulting the right hand column of Table 1.1, SIL 2 is applicable.

This is an example of a **low demand** safety-related system in that it is only called upon to operate at a frequency determined by the frequency of failure of the equipment under control (EUC) – in this case 2×10^{-1} pa. Note, also, that the target 'E' in the above paragraph is dimensionless by virtue of dividing a rate by a rate. Again, this is consistent with the right hand column of Table 1.1 in Chapter 1.

Simple example (high demand)

Now consider an example where a failure in a domestic appliance leads to overheating and subsequent fire. Assume, again, that the target risk of fatality is said to be 10^{-5} pa. Assume that a study suggests that 1 in 400 incidents leads to fatality.

It follows that the target maximum tolerable failure rate for the hazardous event can be calculated as $10^{-5} \times 400 = 4 \times 10^{-3}$ pa (i.e. 1/250 years). This is 4.6×10^{-7} per hr when expressed in units of "per hour" for the purpose of Table 1.1.

Consulting the middle column of Table 1.1, SIL 2 is applicable. This is an example of a **high demand** safety-related system in that it is "at risk" continuously. Note, also, that the target in the above paragraph has the dimension of rate by virtue of multiplying a rate by a dimensionless number. Again, this is consistent with the middle column of Table 1.1.

It is worth noting that for a low demand system the Standard, in general, is being applied to an "add-on" safety system which is separate from the normal control of the EUC (i.e. plant). On the other hand for a continuous system the Standard, in general, is being applied to the actual control element because its failure will lead directly to the potential hazard even though the control element may require additional features to meet the required integrity. Remember (as mentioned in Chapter 1) that a safety-related system with a demand rate of greater than once per annum should be treated as "high demand".

More complex example

In the Fault Tree (Figure 2.1), Gate G1 describes the causes of some hazardous event. It would be quantified using the rate parameter. Dividing the target Maximum Tolerable Failure Rate associated with the top gate (GTOP) by the rate for Gate G1 provides a target PFD (Probability of failure on demand) for the protection.

Independent levels of protection are then modeled as shown by gates G21 and G22 in Figure 2.1. It is important to remember that the use of an AND gate (e.g. Gate G2) implies that the events below that gate are totally independent of each other.

A greater number of levels of protection (i.e. Gates below G2) leads to larger PFDs being allocated for each and, thus, lower integrity requirements will apply to each.

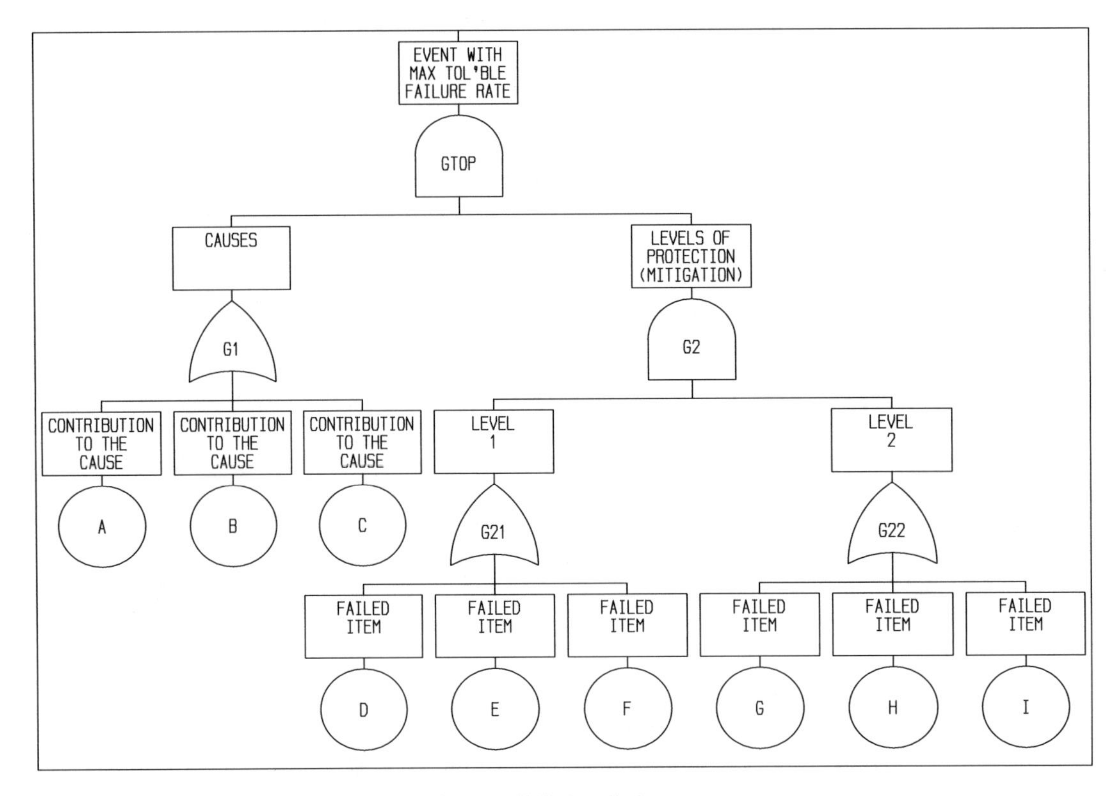

Figure 2.1: Fault Tree.

Table 2.5: Possible SIL outcomes.

	Level 1 PFD	Level 1 SIL	Level 2 PFD	Level 2 SIL
Option	2×10^{-1}	<1	2.65×10^{-2}	1
Option	7.3×10^{-2}	1	7.3×10^{-2}	1
Option	7×10^{-1}	<1	7.57×10^{-3}	2

A maximum tolerable failure rate of 5.3×10^{-4} pa is taken as an example. Assume that the frequency of causes (i.e. Gate G1) is 10^{-1} pa. Thus the target PFD associated with Gate G2 becomes:

$$5.3 \times 10^{-4}\ \text{pa}/\ 10^{-1}\ \text{pa} = 5.3 \times 10^{-3}$$

(Note that the result is dimensionally correct, i.e. a rate/rate becomes a PFD.)

A common mistake is to describe the scenario as "a SIL 2 safety system". This would ONLY be the case if the mitigation were to be a single element and not decomposed into separate independent layers.

In Figure 2.1 there are 2 levels of protection for which the product of the 2 PFDs needs to be less than 5.3×10^{-3}.

Depending on the equipment in question this could involve a number of possibilities. Examples are shown in Table 2.5, which assume independent levels of protection.

As can be seen, the safety integrity level is inferred only once the PFD associated with each level of protection has been assigned/assessed.

(d) Exercises

Now try the following exercises (answers in Appendix 5), which involve establishing SIL targets:

Exercise 1:

Assume a maximum tolerable risk target of 10^{-5} pa (Public fatality)

Assume 1 in 2 Incidents Lead To An Explosion

Assume 1 in 5 explosions lead to a fatality

Assume that a fault tree indicates that the process will suffer a failure rate of 0.05 pa

It is proposed to implement an add-on safety system involving instrumentation and shut-down measures

Which type of SIL (high/low) is indicated and why?

What is the target and what SIL is inferred?

Exercise 2:

2.1

Assume a maximum tolerable risk fatality target of 10^{-5} pa

Assume that there are 9 other similar toxic spill hazards to be assessed from the plant which will threaten the same group of people at the same time

Assume toxic spillage causes fatality 1 in 10 times

Assume that a fault tree indicates that each of the processes will suffer an incident once in 50 years

It is proposed to implement an add-on safety system with instrumentation and shut-down measures

Which type of SIL is indicated and why?

What is the target and what SIL is inferred?

2.2

If additional fire fighting equipment were made available, to reduce the likelihood of a fatality from 1 in 10 to 1 in 30, what effect, if any, is there on the target SIL?

Exercises 1 and 2 involved the low demand table in which the risk criteria were expressed as a probability of failure on demand (PFD). Now try Exercise 3.

Exercise 3:

Target maximum tolerable risk = 10^{-5} pa

Assume that 1 in 200 failures, whereby an interruptible gas meter spuriously closes and then opens, leads to fatality

Which type of SIL is indicated and why?

What is the target and what SIL is inferred?

A point worth pondering is that when a high demand SR system fails, continued use is usually impossible, whereas, for the low demand system, limited operation may still be feasible after the risk reduction system has failed, albeit with additional care.

2.1.2 LOPA (Levels of Protection Analysis)

A methodology, specifically mentioned in Part 3 of IEC 61511 (Annex F), is known as Layer of Protection Analysis (LOPA). LOPA provides a structured risk analysis that can follow on from a qualitative technique such as HAZOP.

In general, formalized LOPA procedures tend to use order of magnitude estimates and are thus referred to as so-called **semi-quantitative** methods. Also, they are tailored to low demand safety functions.

Nevertheless, many practitioners, despite using the term LOPA, actually carry out the analysis to a refinement level such as we have described in section 2.1.1. This is commonly referred to as a **quantitative** approach.

LOPA estimates the probability/frequency of the undesired consequence of failure by multiplying the frequency of initiating events by the product of the probabilities of failure for the applicable protection layers. The severity of the consequences and the likelihood of occurrence are then assigned a probability (often by reference to a standard table usually specified in the user's procedure).

The result is called a "mitigated consequence frequency" and is often compared to a Company's tolerable risk criteria (e.g. Personnel, Environment, Asset Loss). As a result any requirement for additional risk reduction required is identified. The output of the LOPA analysis is the target PFD for the safety instrumented function.

For the LOPA to be valid there must be independence between initiating events and layers of protection and between the layers of protection. Where there are common causes either a dependent layer should not be credited at all or reduced credit (higher PFD) used.

It should also be noted that the maximum tolerable risk frequencies used are usually for ALL hazards. Thus where personnel are exposed to multiple simultaneous hazards, the maximum tolerable risk frequency needs to be divided by the number of hazards.

The input information required for a LOPA includes:

- Process plant and equipment design specifications
- Impact event descriptions and consequence of failure (assuming no protection)
- Severity level category (defined in the Company's procedure)
- All potential demands (i.e. initiating causes) on the function; and corresponding initiation likelihood
- Vulnerability (e.g. probability of a leakage leading to ignition)
- Description of the safety instrumented protection function (i.e. layer of protection)
- Independent Protection Layers (e.g. mechanical devices, physical bunds).

LOPA Worksheets are then prepared as shown in the example given in Chapter 13.6 and are not unlike Table 2.4 and its associated examples. Elements in the worksheet include:

Consequence: describes the consequence of the hazard corresponding to the descriptions given in the user's procedure.
Maximum Tolerable Risk (/yr): as specified in the user's procedure

Lists the identified causes of the hazard.

Initiating Likelihood (/yr): quantifies the expected rate of occurrence of the initiating cause. This rate is based on the experience of the attendees and any historical information available.

Vulnerability: this represents the probability of being affected by the hazard once it has been initiated.

Independent Protection Layers: the level of protection provided by each IPL is quantified by the probability that it will fail to perform its function on demand. The smaller the value of the PFD, the larger the risk reduction factor that is applied to the calculated initiating likelihood [0], hence where no IPL is claimed, a '1' is inserted into the LOPA worksheet.

The outputs from a LOPA include:

- Intermediate event likelihoods (assuming no additional instrumented protection);
- Additional protection instrumentation requirements (if any);
- The mitigated event likelihood.

2.1.3 The Risk Graph Approach

In general the methods described in sections 2.1.1 and 2.1.2 should be followed. However, the Standard acknowledges that a fully quantified approach to setting SIL targets is not always possible and that an alternative approach might sometimes be appropriate. This avoids quantifying the maximum tolerable risk of fatality and uses semi-quantitative judgements. Figure 2.2 gives an example of a risk graph.

The example shown is somewhat more complete than many in use. It has the additional granularity of offering 3 (rather than 2) branches in some places and attempts to combine demand rate with exposure. Any such approach requires a detailed description of the decision points in the algorithm in order to establish some conformity of use. Table 2.6 shows a small part of that process.

Risk graphs should only be used for general guidance in view of the wide risk ranges of the parameters in the tables. Successive cascading decisions involving only "order of magnitude" choices, carry the potential for gross inaccuracy. Figure 2.2 improves on the granularity which simple risk graphs do not offer. Nevertheless this does not eliminate the problem.

The risk graph does not readily admit multiple levels of protection. This has been dealt with in earlier sections. Furthermore, due to the nature of the rule based algorithm, which culminates in the request for a demand rate, the risk graph is only applicable to low demand SIL targets. It should only be used as a screening tool when addressing large numbers of safety functions. Then, any target of SIL 2 or greater should be subject to the quantified approach.

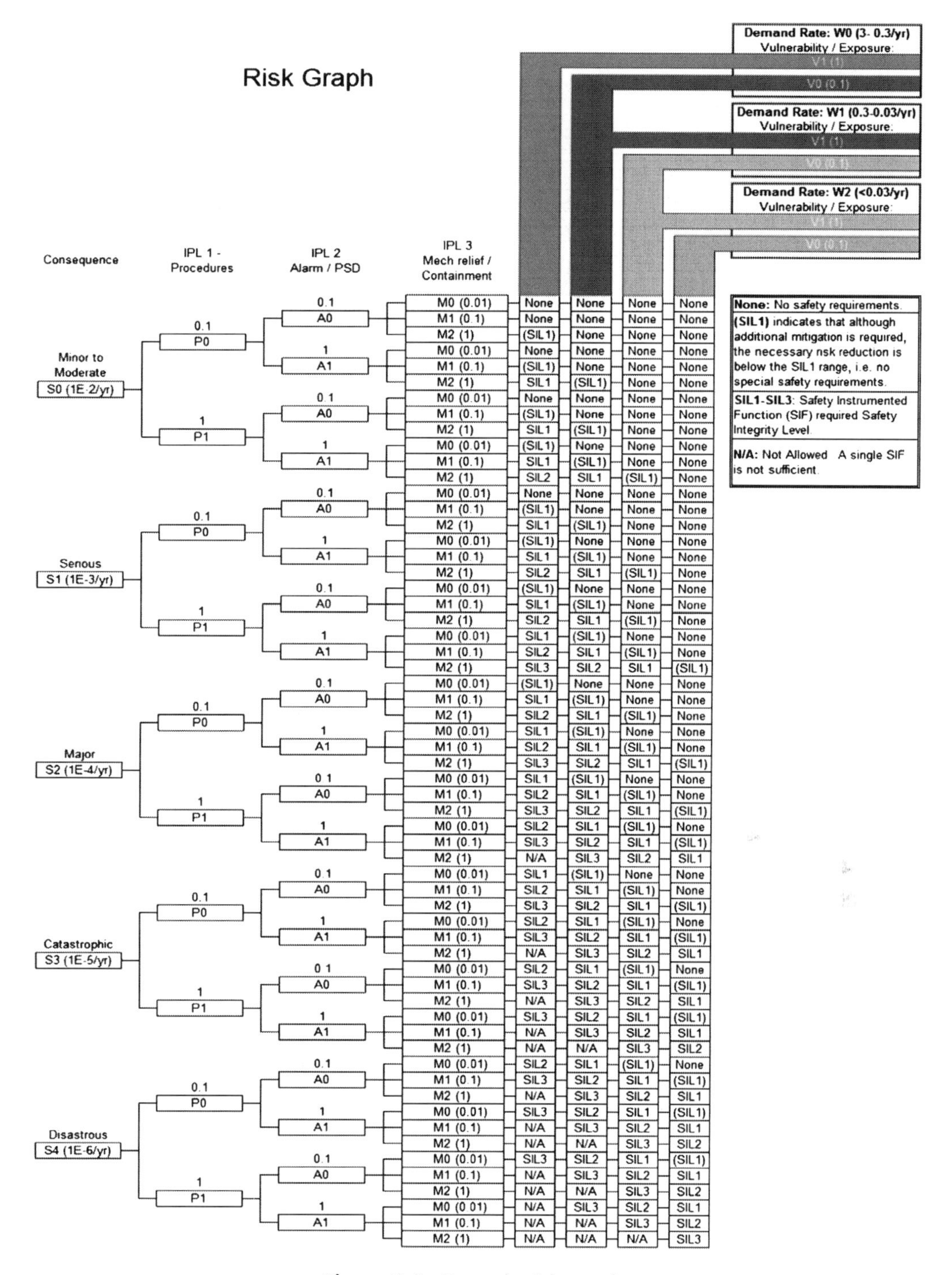

Figure 2.2: Example risk graph.

Table 2.6: Key to Figure 2.1 (part of only).

Major - permanent effect		**S0**	
Catastrophic - 1 fatality and/or several permanent disabilities		**S1**	
Independent Protection Layers		**Cat.**	**Value**
Procedures	Clear, documented procedure in place.	P0	0.1
	None	P1	1
Alarm / PSD	Independent alarm / PSD in place	A0	0.1
	None	A1	1
Mechanical relief / containment	2 mechanical relief / containment devices in place	M0	0.01
	1 mechanical relief / containment devices in place	M1	0.1
	None	M2	1

2.1.4 *Safety Functions*

IMPORTANT: It should be clear from the foregoing sections that SILs are ONLY appropriate to specifically defined safety functions. A safety function might consist of a flow transmitter, logic element and a solenoid valve to protect against high flow. The flow transmitter, on its own, does not have a SIL and to suggest such is nearly meaningless. Its target SIL may vary from one application to another. The only way in which it can claim any SIL status in its own right is in respect of safe failure fraction and of the life-cycle activities during its design, and this will be dealt with in Chapters 3 and 4.

2.1.5 *"Not Safety-Related"*

It may well be the case that the SIL assessment indicates a probability or rate of failure less than is indicated for SIL 1. In this case the system may be described as "not safety-related" in the sense of the Standard. However, since the qualitative requirements of SIL 1 are little more than established engineering practice they should be regarded as a "good practice" target.

The following example shows how a piece of control equipment might be justified to be "not safety-related". Assume that this programmable Distributed Control System (say a DCS for a process plant) causes various process shutdown functions to occur. In addition, let there be a hardwired Emergency Shutdown (presumably safety-related) system which can also independently bring about these shutdown conditions.

Assume the target maximum tolerable risk leads us to calculate that the failure rate for the DCS/ESD combined should be better than 10^{-3} pa. Assessment of the emergency shutdown system shows that it will fail with a PFD of 5×10^{-3}. Thus, the target maximum tolerable failure rate of the DCS becomes 10^{-3} pa/ $5 \times 10^{-3} = 2 \times 10^{-1}$ pa. This being less onerous than the target for SIL 1, the target for the DCS is less than SIL 1. This is ambiguously referred to as "not safety-related". An alternative term used in some guidance documents is "no special safety requirement".

We would therefore say that the DCS is not safety-related. If, on the other hand, the target was only met by a combination of the DCS and ESD then each might be safety-related with a SIL appropriate to its target PFD or failure rate. Paragraph 7.5.2.5 of Part 1 states that the EUC must be $<$ SIL 1 or else it must be treated as safety-related.

For less than SIL 1 targets, the term SIL 0 (although not used in the Standard) is in common use and is considered appropriate.

2.1.6 SIL 4

There is a considerable body of opinion that SIL 4 safety functions should be avoided (as achieving it requires very significant levels of design effort and analysis) and that additional levels of risk reduction need to be introduced such that lower SIL targets are required for each element of the system.

In any case, a system with a SIL 4 target would imply a scenario with a high probability of the hazard leading to fatality and only one level of control (i.e. no separate mitigation). It is hard to imagine such a scenario as being acceptable.

2.1.7 Environment and Loss of Production

So far the implication has been that safety-integrity is in respect of failures leading to death or injury. IEC 61508 (and some other guidance) also refers to severe environmental damage. Some guidance documents provide a risk graph for establishing a target SIL for equipment where failure leads to such an outcome (Figure 2.3). It is not known how the Figure 2.3 algorithm was developed.

Furthermore, although not directly relevant here, the same SIL approach can be applied to loss of production.

An alternative approach would be to establish a "maximum acceptable annual cost". Then, the probability of failure on demand might be assessed as the ratio:

$$\frac{\text{"Maximum acceptable annual cost"}}{(\text{Cost of the Consequence} \times \text{Frequency of occurrence})}$$

Consequence severity	Demand rate: Low	Demand rate: Verylow	Demand rate: Relatively high
No release or a negligible environmental impact	1		
Release with minor impact on the environment	2	1	
Release with moderate impact on the environment	3	3	2
Release with temporary major impact on environ't	NO	NO	3
Release with permanent major impact on environ't			

Figure 2.3: Environmental risk graph.

The PFD could then be translated into a SIL using the low demand table.

2.1.8 Malevolence and Misuse

Paragraph 7.4.2.3 of part 1 of the standard

The 2010 version of IEC 61508 draws attention to the need to address all foreseeable causes of a hazard. Thus human factors (already commonly addressed) should be extended to include vandalism, deliberate misuse, criminal interference and so on. The frequency of such events can be assessed (anecdotally or from records) enabling them to be included in fault tree models.

2.2 ALARP ("As low as Reasonably Practicable")

Having established a SIL target it is insufficient merely to assess that the design will meet the maximum tolerable risk target. It is necessary to establish whether further improvements are justified and thus the principle of ALARP (as low as reasonably practicable) is called for as "good practice". In the UK this is also arguably necessary in order to meet safety legislation ("all that is reasonably practicable" is called for in the Health & Safety at Work Act 1974).

Figure 2.4 shows the so called ALARP triangle which also makes use of the idea of a Maximum Tolerable Risk.

In this context "acceptable" is generally taken to mean that we accept the probability of fatality as being reasonably low, having regard to the circumstances, and would not usually seek to expend more resources in reducing it further.

"Tolerable", on the other hand, implies that whilst we are prepared to live with the particular risk level we would continue to review its causes and the defenses we might take with a view to

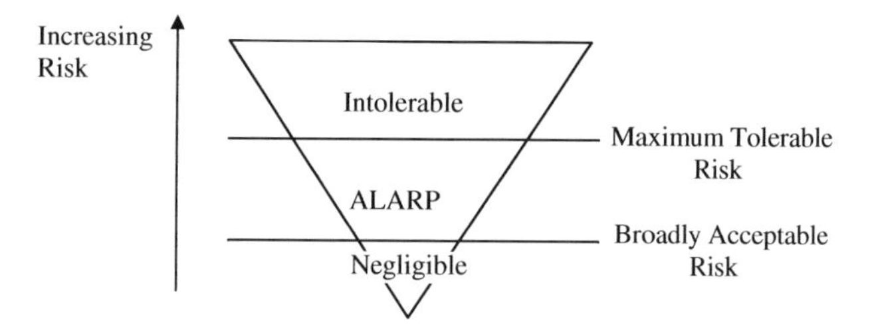

Figure 2.4: ALARP triangle.

reducing it further. Cost comes into the picture in that any potential reduction in risk would be compared with the cost needed to achieve it.

"Unacceptable" means that we would not normally tolerate that level of risk and would not participate in the activity in question nor permit others to operate a process that exhibited it except, perhaps, in exceptional circumstances.

The principle of ALARP (as low as reasonably practicable) describes the way in which risk is treated legally and by the HSE in the UK, and also applied in some other countries. The concept is that all reasonable measures will be taken in respect of risks which lie in the tolerable (ALARP) zone to reduce them further until the cost of further risk reduction is grossly disproportionate to the benefit.

It is at this point that the concept of "cost per life saved" arises. Industries and organizations are reluctant to state specific levels of "cost per life saved" which they would regard as being disproportionate to a reduction in risk. However, criteria in the range £1,000,000 to £15,000,000 are not infrequently quoted.

Perception of risk is certainly influenced by the circumstances. A far higher risk is tolerated from voluntary activities than from involuntary ones (people feel that they are more in control of the situation on roads than on a railway). This explains the use of different targets for employee (voluntary) and public (involuntary) in Tables 2.1 – 2.3.

A typical ALARP calculation might be as follows:

A £1,000,000 cost per life saved target is used in a particular industry.
A maximum tolerable risk target of 10^{-4} pa has been set for a particular hazard which is likely to cause 2 fatalities.
The proposed system has been assessed and a predicted risk of 8×10^{-5} pa obtained.
Given that the negligible risk is taken as 10^{-6} pa then the application of ALARP is required.
For a cost of £3,000, additional instrumentation and redundancy will reduce the risk to just above the negligible region (2×10^{-6} pa).
The plant life is 30 years.

Hence cost per life saved $= £3,000/(8 \times 10 - 5 - 2 \times 10 - 6) \times 2 \times 30 = £640,000$

This being less than the £1,000,000 cost per life saved criterion the proposal should be adopted. It should be noted that all the financial benefits of the proposed risk reduction measures should be included in the cost benefit calculation (e.g. saving plant damage, loss of production, business interruption etc.). Furthermore, following "good practice" is also important although not of itself sufficient to demonstrate ALARP. Cost—benefit arguments should not be used to justify circumventing established good practice.

Exercise 4:

A £2,000,000 cost per life saved target is used in a particular industry

A maximum tolerable risk target of 10^{-5} pa has been set for a particular hazard which is likely to cause 3 fatalities

The proposed system has been assessed and a predicted risk of 8×10^{-6} pa obtained

How much could justifiably be spent on additional instrumentation and redundancy to reduce the risk from 8×10^{-6} pa to 2×10^{-6} pa (just above the negligible region)?

The plant life is 25 years

In order to demonstrate that ALARP has been achieved, it is necessary to show that the cost of implementing a measure to reduce risk is grossly disproportionate to the benefit. There are no hard and fast rules; however, some guidance is given in the HSE documents HSE SPC/Permissioning/9 and HSE SPC/Permissioning/12.

The suggestion is that the cost per life saved criterion is multiplied by a gross disproportion factor of between 1 and 2 towards the bottom of the ALARP region (i.e. just above the "broadly acceptable" level) and 10 towards the top of the ALARP region (i.e. just below the "intolerable" level).

If the risk is in the "intolerable" region, then risk reduction measures must be implemented irrespective of the cost.

2.3 Functional Safety Management and Competence

2.3.1 Functional Safety Capability Assessment

In claiming conformance (irrespective of the target SIL) it is necessary to show that the management of the design, operations and maintenance activities and of the system implementation is itself appropriate and that there is adequate competence for carrying out each task.

This involves two basic types of assessment. The first is the assessment of management procedures (similar to but more rigorous than an ISO 9001 audit). Appendix 1 of this book provides a Functional Safety Capability template procedure which should be adequate as an addition to an ISO 9001 quality management system. The second is an assessment of the implementation of these procedures. Thus, the life-cycle activities described in Chapters 1, 3 and 4 would be audited, for one or more projects, to establish that the procedures are being put into practice.

Appendix 2 contains a checklist schedule to assist in the rigor of assessment, particularly for self assessment (see also Chapter 7.3).

2.3.2 Competency

In Part 1 of IEC 61508 (Paragraphs 6.2.13–15) the need for adequate competency is called for. It is open-ended in that it only calls for the training, knowledge, experience and qualifications to be "relevant". Factors listed for consideration are:

- Responsibilities and level of supervision
- Link between severity of consequences and degree of competence
- Link target SIL and degree of competence
- The link between design novelty and rigor of competence
- Relevance of previous experience
- Engineering application knowledge
- Technology knowledge
- Safety engineering knowledge
- Legal/regulatory knowledge
- Relevance of qualifications
- The need for training to be documented.

(a) IET/BCS "Competency guidelines for safety-related systems practitioners"

This was an early guidance document in this area. It listed 12 safety-related job functions (described as functions) broken down into specific tasks. Guidance is then provided on setting up a review process and in assessing capability (having regard to applications relevance) against the interpretations given in the document. The 12 jobs are:

Corporate Functional Safety Management: This concerns the competency required to develop and administer this function within an organization.
Project Safety Assurance Management: This extends the previous task into implementing the functional safety requirements in a project.
Safety-Related System Maintenance: This involves maintaining a system and controlling modifications so as to maintain the safety-integrity targets.

Safety-related System Procurement: This covers the technical aspects of controlling procurement and sub-contracts (not just administration).
Independent Safety Assessment: This is supervising and/or carrying out the assessments.
Safety Hazard and Risk Analysis: That is to say HAZOP (HAZard and OPerability study), LOPA, risk analysis, prediction etc.
Safety Requirements Specification: Being able to specify all the safety requirements for a system.
Safety Validation: Defining a test/validation plan, executing and assessing the results of tests.
Safety-related System Architectural Design: Being able to partition requirements into sub-systems so that the overall system meets the safety targets.
Safety-related System Hardware Realization: Specifying hardware and its tests.
Safety-related System Software Realization: Specifying software, developing code and testing the software.
Human Factors Safety Engineering: Assessing human error and engineering the inter-relationships of the design with the human factors (Chapter 5.4)

The three levels of competence described in the document are:

The Supervised Practitioner who can carry out one of the above jobs but requiring review of the work.
The Practitioner who can work unsupervised and can manage and check the work of a Supervised Practitioner.
The Expert who will be keeping abreast of the state of art and will be able to tackle novel scenarios.

This IET/BCS document provided a solid basis for the development of competence. It probably goes beyond what is actually called for in IEC 61508. Due to its complexity it is generally difficult to put into practice in full and therefore might discourage some people from starting a scheme. Hence a simpler approach might be more practical. However, this is a steadily developing field and the requirements of "good practice" are moving forward.

(b) HSE document (2007) "Managing competence for safety-related systems"

More recently, this document was produced in co-operation with the IET and the BCS. In outline its structure is:

Phase One – Plan
    Define purpose and scope
Phase Two – Design
    Competence criteria
    Processes and methods

Phase Three – Operate
- Select and recruit
- Assess competence
- Develop competence
- Assign responsibilities
- Monitor
- Deal with failure
- Manage assessors' and managers' competence
- Manage supplier competence
- Manage information
- Manage change

Phase Four – Audit and Review
- Audit
- Review

(c) Annex D of "Guide to the application of IEC61511"

This is a fairly succinct summary of a competency management system which lists competency criteria for each of the life-cycle phases described in Chapter 1.4 of this book.

(d) Competency register

Experience and training should be logged so that individuals can be assessed for the suitability to carry out tasks as defined in the company's procedure (Appendix 1 of this book).

Figure 2.5 shows a typical format for an Assessment Document for each person. These would form the competency register within the organization.

2.3.3 Independence of the Assessment

This is addressed in Part 1 - 8.2.18. The level of independence to be applied when carrying out assessments is recommended, according to the target SIL, can be summarized as:

SIL	Assessed by:
4	Independent organization
3	Independent department
2	Independent person
1	Independent person

For SILs 2 and 3 add one level of independence if there is lack of experience, unusual complexity or novelty of design. Clearly, these terms are open to interpretation and words such

Name	Xxxxxxx	
Qualifications	BSc, MSc in Safety (xx University)	
Date of employment	Xxxxxxx	
Training	In-house appreciation course Technis certificate in R&FS (distinction)	May 2008 April 2010
Professional	Paper on QRA and maximum tolerable risk comparisons (SaRS Journal)	2009
Task in the Life-cycle	Experience	Level of expertise (as defined in company procedure)
Risk Analysis	Lead SIL determination team 5 processes (2009)	FS Manager
Requirements	Reviewed requ's specs for new instrumentation (ESD and HIPPs systems) and drafted FS requ's	FS Assessor
Design	No experience to date	N/A
Assessment	Introduced Fault Tree tool and carried out 6 assessments of ESD systems against SIL targets. Analysed field data over a 3 year period and produced failure rate sheet for instruments and actuators	FS Manager
Regulatory	Attended 3 meetings with HSE representatives: a) Review of human factors elements of company safety submissions b) Review of SIL targets c) Review of life-cycle claims	FS Assessor
etc	etc	
etc	etc	
Training Needs	Design of ESD architectures and choice of instrumentation to meet SIL targets Review of life cycle techniques and measures	
Last Review	31 May 2010 by xxx and yyy	

Figure 2.5: Competency register entry.

as "department" and "organization" will depend on the size and type of company. For example, in a large multi-project design company there might be a separate safety assessment department sufficient to meet the requirements of SIL 3. A smaller single-project company might, on the other hand, need to engage an independent organization or consultant in order to meet the SIL 3 requirement.

The level of independence to be applied when establishing SIL targets is recommended, according to consequence, as:

Multiple fatality, say >5	Independent organization
Multiple fatality	Independent department
Single fatality	Independent person
Injury	Independent person

For scenarios involving fatality, add one level of independence if there is lack of experience, unusual complexity or novelty of design. Clearly, these terms are open to interpretation and words such as "department" and "organization" will depend on the size and type of company.

2.3.4 Hierarchy of Documents

This will vary according to the nature of the product or project and the life-cycle activities involved. The following brief outline provides an overview from which some (or all) of the relevant documents can be taken.

Annex A of Part 1 addresses these lists. The following is an interpretation of how they might be implemented. It should be stressed that document titles (in themselves) need not be rigidly adhered to and that some might be incorporated into other existing documents. An example is the "safety requirements" which might in some cases sit within the "functional specification" providing that they are clearly identified as a coherent section.

- Functional Safety Requirements
- Functional Safety Plan (See Appendix 7 of this book)
- Validation Plan (and report)
- Functional safety design specification (Hardware)
- Functional safety design specification (Software)
- Review Plans (and reports)
- Test Plans (and reports)
- Test strategy and procedures
- Safety Manual (maybe part of Users' Manual).

These are dealt with, as they occur, in Chapters 3 and 4.

2.3.5 Conformance Demonstration Template

In order to justify adequate functional safety management to satisfy Part 1 of the standard, it is necessary to provide a documented assessment.

The following Conformance Demonstration Template is suggested as a possible format.

IEC 61508 Part 1

Under "Evidence" enter a reference to the project document (e.g. spec, test report, review, calculation) which satisfies that requirement. Under "Feature" read the text in conjunction with the fuller text in this chapter.

Feature	Evidence
Adequate Functional Safety Capability is demonstrated by the organization. To include a top level policy, clear safety life-cycle describing the activities undertaken, procedures, functional safety audits and arrangements for independent assessment	
FS management system regularly reviewed and audited	
An adequate competency register which maps to projects and the requirement for named individuals for each FS role. Register to describe training and application area experience of individuals. Safety related tasks to be defined. Review and training to be covered	
Evidence that contract and project reviews are mandatory to establish functional safety requirements	
The need for a clear documentation hierarchy describing the relationship of Q&S Plan, Functional Spec, Design docs, Review strategy, Integration, Test and Validation plans etc.	
Existence of hardware and software design standards and defined hardware and software life-cycle models	
The recording and follow-up of hazardous incidents. Adequate corrective action	
Hazardous incidents addressed and handled	
Operations and maintenance adequately addressed where relevant	
Modifications and impact analysis addressed and appropriate change documentation	
Document and configuration adequate control	
FS Assessment carried out	

It is anticipated that the foregoing items will be adequately dealt with by the organization's quality managements systems and the additional functional safety procedure exampled in Appendix 1 of this book.

CHAPTER 3

Meeting IEC 61508 Part 2

Chapter Outline

Safety Critical Systems Handbook. DOI: 10.1016/B978-0-08-096781-3.10003-3

IEC 61508 Part 2 covers the safety system hardware and overall system design, whereas software design is covered by Part 3 (see next Chapter). This chapter summarizes the main requirements. However, the following points should be noted first.

The appropriateness of each technique, and the degree of refinement (e.g. high medium low), represents the opinions of individuals involved in drafting the Standard.

The combination of text (e.g. paras 7.1 to 7.9) and tables (both A and B series) and the use of modifying terms (such as high, medium and low) to describe the intensity of each technique has led to a highly complex set of requirements. Their interpretation requires the simultaneous reading of textual paragraphs, A tables, B tables and Table B6 – all on different pages of the standard. The A Tables are described as referring to measures for controlling (i.e. revealing) failures and the B Tables to avoidance measures.

The authors of this book have, therefore, attempted to simplify this "algorithm of requirements" and this Chapter is offered as a credible representation of requirements.

At the end of this Chapter a "conformance demonstration template" is suggested which, when completed for a specific product or system assessment, will offer evidence of conformance to the SIL in question.

The approach to the assessment will differ substantially between:

COMPONENT (e.g. Transducer) DESIGN

and

APPLICATIONS SYSTEM DESIGN

The demonstration template tables at the end of this chapter cater for the latter case. Chapter 8, which covers the restricted subset of IEC 61511, also caters for applications software.

3.1 Organizing and Managing the Life-cycle

Sections 7.1 of the Standard: Table '1'

The idea of a design life-cycle has already been introduced to embrace all the activities during design, manufacture, installation and so on. The exact nature of the design-cycle model will depend on complexity and the type of system being designed. The IEC 61508 model (in Part 1 of the Standard) may well be suitable and was fully described in Chapter 1 of this book. In IEC 61508 Part 2 its Table '1' describes the life-cycle activities again and is, more or less, a repeat of Part 1.

A major point worth making is that the life-cycle activities should all be documented. Unless this is done, there is no visibility to the design process and an assessment cannot verify that the standard has been followed. This should be a familiar discipline in as much as most readers will be operating within an ISO 9001 management system of practice. The design should be conducted under a project management regime and adequately documented to provide traceability. These requirements can be met by following a quality system such as specified in ISO 9001. The level and depth of the required project management and documentation will depend on the SIL level. The use of checklists is desirable at all stages.

The need for Functional Safety Capability (more recently called Functional Safety Management) has been described in Chapter 2, section 2.3 and also in Appendix 1. IEC 61508 Part 2 (Hardware) and Part 3 (Software) expect this to have been addressed.

Irrespective of the target SIL there needs to be a project management structure which defines all the required actions and responsibilities, along with defining adequate competency, of the persons responsible for each task. There needs to be a "Quality and Safety" Plan which heads the documentation hierarchy and describes the overall functional safety targets and plans. All documentation and procedures need to be well-structured, for each design phase, and sufficiently clear that the recipient for the next phase can easily understand the inputs to that task. This is sufficiently important that Appendix 7 of this book provides more detail.

SIL 3 and SIL 4 require, also, that the project management identify the additional procedures and activities required at these levels and that there is a robust reporting mechanism to confirm both the completion and correctness of each activity. The documentation used for these higher SIL systems should be generated based on standards which give guidance on consistency and layout and include checklists. In addition, for SIL 4 systems, computer aided configuration control and computer aided design documentation should be used. Table B6 of the Standard elaborates on what constitutes a higher rigor of techniques and measures. Project Management, for example, requires validation independent from design and using a formalized procedure, computer aided engineering etc in order to attract the description "high effectiveness".

Much of the above "good practice" (e.g. references to Project Management) tends to be repeated, throughout the Standard, for each of the life-cycle activities, in both text and Tables. We have attempted to avoid such repetition in this book. There are many other aspects of the Standard's guidance which are repetitious and we have tended to refer to each item once and in the most appropriate Section.

The need for validation planning is stressed in the Standard and this should be visible in the project Quality/Safety Plan which will include reference to the Functional Safety Audits.

In general this whole section should be met by implementing the template Functional Safety Procedure provided in Appendix 1.

3.2 Requirements Involving the Specification

Section 7.2 of the Standard: Table B1 [avoidance]

(a) The safety requirements specification

This is an important document because it is crucial to specify the requirements of a safety system correctly and completely. Irrespective of the SIL target it should be clear, precise, unambiguous, testable and well structured, and cover:

- Description of the hazards
- Integrity level requirements plus type of operation, i.e. low demand or high demand for each function
- Response times
- Safety function requirements, definition of the safe state and how it is achieved
- System documents (e.g. P&IDS, Cause and Effect matrices, logic diagrams, process data sheets, equipment layouts)
- System architecture
- Operational performance and modes of operation
- Behavior under fault conditions
- Start-up and re-set requirements
- Input ranges and trip values, outputs, over-rides
- Manual shut-down details
- Behavior under power loss
- Interfaces with other systems and operators
- Environmental design requirements for the safety system equipment
- Electro-magnetic compatibility
- Requirements for periodic tests and/or replacements
- Separation of functions (see below)
- Deliverables at each life-cycle stage (e.g. test procedures, results).

Structured design should be used at all SIL levels. At the system application level the functional requirements (i.e. logic) can be expressed by using semi-formal methods such as cause and effect diagrams or logic/function block diagrams. All this can be suitable up to SIL 3. These include Yourdon, MASCOT, SADT, and several other techniques referenced in Part 7 of the Standard. In the case of new product design rather than applications engineering (i.e. design of executive software) structured methods should be progressively considered from SIL 2 upwards. For SIL 4 applications structured methods should be used

(b) Separation of functions

In order to reduce the likelihood of common cause failures the specification should also cover the degree of separation required, both physically and electrically, between the EUC and the safety system(s). Any necessary data interchange between the two systems should also be tightly specified and only data flow *from* the EUC *to* the safety system permitted.

These requirements need to be applied to any redundant elements of the safety-related system(s).

Achieving this separation may not always be possible since parts of the EUC may include a safety function that cannot be dissociated from the control of the equipment. This is more likely for the continuous mode of operation in which case the whole control system should be treated as safety-related pending target SIL calculations (Chapter 2, section 2.1).

If the safety-related and non-safety-related system elements cannot be shown to be sufficiently independent then the complete system should be treated as safety-related.

For SIL 1 and SIL 2 there should be a clear specification of the separation between the EUC and the safety system and electrical/data interfaces should be well defined. Physical separation should be considered.

For SIL 3 there should be physical separation between the EUC and the safety system and, also, the electrical/data interfaces should be clearly specified. Physical separation of redundant parts of the safety system should be considered.

For SIL 4 there should be total physical/electrical/data separation between the safety system and the EUC and between the redundant parts of the safety system.

3.3 Requirements for Design and Development

Section 7.4 of the Standard: Table B2 [avoidance]

3.3.1 Features of the Design

Sections 7.4.1–7.4.11 excluding 7.4.4 and 7.4.5

(a) Use of 'in-house' design standards and work practices needs to be evident. These will address proven components and parts, preferred designs and configurations etc.

(b) On manual or auto-detection of a failure the design should ensure system behavior which maintains the overall safety targets. In general, this requires that failure in a safety system having redundant paths should be repaired within the mean time to repair assumed in the hardware reliability calculations. If this is not possible, then the procedure should be the same as for non-redundant paths as follows. On failure of a safety system with no redundant paths, either additional process monitoring should be provided to maintain adequate safety or the EUC should be shut down.

(c) Sector specific requirements need to be observed. Many of these are contained in the documents described in Chapters 8-10.

(d) The system design should be structured and modular and use well-tried modules/components. Structured, in this context, implies clear partitioning of functions and a visible hierarchy of modules and their interconnection. For SIL 1 and SIL 2 the modularity should be kept to a "limited size" and each module/component should have had previously documented field experience for at least one year with 10 devices. If previous experience does not exist, or is insufficiently documented, then this can be replaced with additional modular/component testing. Such use of subjective descriptions (e.g. the "limited size") adds further weight to the desirability of "in-house" checklists, which can be developed in the light of experience.

In addition for SIL 3 systems, previous experience is needed in a relevant application and for a period of at least two years with ten devices or, alternatively, some third party certification.

SIL 4 systems should be both proven in use, as mentioned above, and have third-party certification.

It is worth mentioning that the "years" of operation referred to above assume full time use (i.e. 8760 hrs per annum).

(e) Systematic failures caused by the design (this refers to Tables A15 and A18): the primary technique is to use monitoring circuitry to check the functionality of the system. The degree of complexity required for this monitoring ranges from "low" for SIL 1 and SIL 2, through "medium" for SIL 3 to "high" for SIL 4.

For example a PLC-based safety system with a SIL 1 or SIL 2 target would require, as a minimum, a watch-dog function on the PLC CPU being the most complex element of this "lower" integrity safety system.

These checks would be extended in order to meet SIL 3 and would include additional testing on the CPU (i.e. memory checks) along with basic checking of the I/O modules, sensors and actuators.

The coverage of these tests would need to be significantly increased for SIL 4 systems. Thus the degree of testing of input and output modules, sensors and actuators would be substantially increased. Again, however, these are subjective statements and standards such as IEC 61508 do not and cannot give totally prescriptive guidance. Nevertheless some guidance is given concerning diagnostic coverage.

It should be noted that the minimum configuration table given in Section 3.3.2a of this chapter permits higher SIL claims, despite lower levels of diagnosis, by virtue of either more redundancy or a higher proportion of "fail safe" type failures. The 2010 version allows a proven-in-use alternative (see 3.3.2b).

(f) Systematic failures caused by environmental stress (this refers to Table A 16): this requirement applies to all SILs and states that all components (indeed the overall system) should be designed and tested as suitable for the environment in question. This includes temperature and temperature cycling, emc (electro-magnetic compatibility), vibration, electro-static, etc. Components and systems that meet the appropriate IEC component standards, or CE marking, UL (Underwriters Laboratories Inc) or FM (Factory Mutual) approval would generally be expected to meet this requirement.

(g) Systematic operation failures (this refers to Table A17): for all SILs the system should have protection against on-line modifications of either software or hardware.

There needs to be feedback on operator actions, particularly when these involve keyboards, in order to assist the operator in detecting mistakes.

As an example of this, for SIL 1 and SIL 2, all input operator actions should be repeated back whereas, for SIL 3 and SIL 4, significant and consistent validation checks should be made on the operator action before acceptance of the commands.

The design should take into account human capabilities and limitations of operators and maintenance staff. Human factors are addressed in Chapter 5.4 of this book.

(h) Tables A1 to A15 of the Standard are techniques considered suitable for achieving improvements in diagnostic capability. The following section 3.3.2 discusses diagnostic capability and SFF. Carrying out a detailed FMEA (Appendix 4) will generally provide a claim of diagnostic capability which over-rides these tables. However, they can be used as a guide to techniques.

(i) Communications: 7.4.11 of the Standard requires one to address the failure rate of the communications process. Channels are described in two ways:

- White Box: where the communications executive software has already been designed and certified to provide the appropriate integrity (e.g. use of self test etc.)
- Black Box: where the integrity is designed in at the applications software level because the whitebox claim cannot be made.

(j) Synthesis of elements: 7.4.3 allows a configuration involving parallel elements, each demonstrating a particular SIL in respect of systematic failures, to claim an increment of one

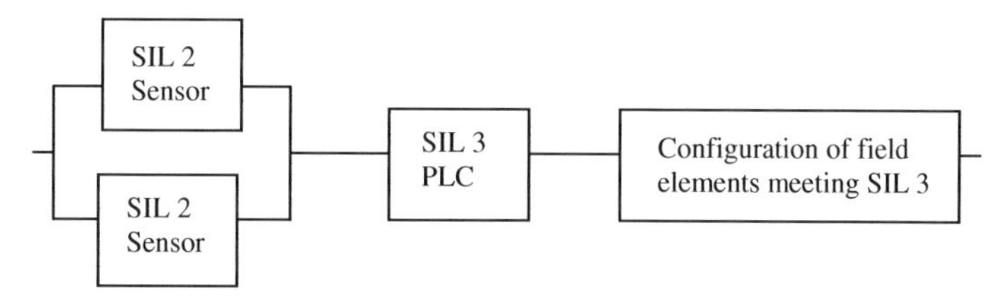

Figure 3.1: Showing two SIL 2 elements achieving a SIL 3 result.

SIL. This requires that a common cause analysis has been carried out in order to demonstrate independence by use of appropriate techniques (e.g. functional diversity). Figure 3.1 illustrates the idea.

3.3.2 Architectures (i.e. Safe Failure Fraction)

Section 7.4.4 Tables '2' and '3'

(a) Claim via SFF (known, in the Standard, as Route 1_H)

Regardless of the hardware reliability calculated for the design, the standard specifies minimum levels of redundancy coupled with given levels of fault tolerance (described by the Safe Failure Fraction).

This Safe Failure Fraction, for each safety function, needs to be estimated as shown in Appendix 4.

The term Safe Failure Fraction (SFF) is coined, in IEC 61508, in addition to the concept of diagnostic coverage. The percentages described as the "safe failure fraction" refer to the sum of the potentially dangerous failures revealed by auto-test together with those which result in a safe state, as a fraction of the TOTAL number of failures. Thus:

$$\text{SFF} = \frac{\text{Total revealed hazardous failures} + \text{Total safe failures}}{\text{Total failures}}$$

("Total failures" are those on the top line PLUS the unrevealed hazardous failures.)

An example might be a slamshut valve where 90% of the failures are "spurious closure" and 10% "fail to close". In that case, a 90% "safe failure fraction" would be claimed without further need to demonstrate automatic diagnosis. On the other hand a combined example might be a control system whereby 50% of failures are "fail-safe" and the remaining 50% enjoy an 80% automatic diagnosis. In this latter case the overall safe failure fraction becomes 90% (i.e. 50% + 0.8 × 50%).

There are two tables which cover the so-called "Type A" components (failure modes well defined PLUS behavior under fault conditions well defined PLUS failure data available) and the "Type B" components (likely to be more complex and whereby any of the above are not satisfied).

In the following Tables "m" refers to the number of failures which lead to system failure. The tables provide the maximum SIL which can be claimed for each safe failure fraction case. The expression "m+1" implies redundancy whereby there are (m+1) elements and m failures are sufficient to cause system failure. The term Hardware Fault Tolerance is commonly used. An HFT of 0 implies simplex (i.e. no failures tolerated). An HFT of 1 implies m out of (m+1) (i.e. 1 failure tolerated) and so on.

Requirements for Safe Failure fraction

Type A SFF	SIL for Simplex HFT 0	SIL for (m+1) HFT 1	SIL for (m+2) HFT 2
<60%	1	2	3
60%-90%	2	3	4
90%-99%	3	4	4
>99%	3	4	4
Type B SFF	**SIL for Simplex HFT 0**	**SIL for (m+1) HFT 1**	**SIL for (m+2) HFT 2**
<60%	NO*	1	2
60%-90%	1	2	3
90%-99%	2	3	4
>99%	3	4	4

Simplex implies no redundancy
(m+1) implies 1 out of 2, 2 out of 3 etc
(m+2) implies 1 out of 3, 2 out of 4 etc
*This configuration is not allowed.

The above table refers to 60%, 90% and 99%. At first this might seem a realistic range of safe fail fraction ranging from simple to comprehensive. However, it is worth considering how the diagnostic part of each of these coverage levels might be established. There are two ways in which diagnostic coverage and safe failure fraction ratios can be assessed:

By test: where failures are simulated and the number of diagnosed failures, or those leading to a safe condition, are counted.
By FMEA: where the circuit is examined to ascertain, for each potential component failure mode, whether it would be revealed by the diagnostic program or lead to a safe condition.

Clearly a 60% safe failure fraction could be demonstrated fairly easily by either method. Test would require a sample of only a few failures to reveal 60%.

Turning to 90% coverage, the test sample would now need to exceed 20 failures (for reasonable statistical significance) and the FMEA would require a more detailed approach. In both cases the cost and time become more significant. An FMEA as illustrated in Appendix 4 is needed and might well involve 3–4 man-days.

For 99% coverage a reasonable sample size would now exceed 200 failures and the test demonstration is likely to be impracticable.

The foregoing should be considered carefully to ensure that there is adequate evidence to claim 90% and an even more careful examination before accepting the credibility of a 99% claim.

In order to take credit for diagnostic coverage, as described in the Standard (i.e. the above Architectural Constraint Tables), the time interval between repeated tests should at least be an order of magnitude less than the expected demand interval. For the case of a continuous system then the auto-test interval plus the time to put the system into a safe state should be within the time it takes for a failure to propagate to the hazard.

Furthermore, it is important to remember that auto-test means just that. Failures discovered by however frequent manual proof-tests are not credited as revealed for the purpose of an SFF claim.

(b) Claim via field failure data (7.4.4.2 of Part 2) (known, in the Standard, as Route 2_H)

The 2010 version of the Standard permits an alternative route to the above "architectures" rules. If well documented and verified FIELD (not warranty/returns) based failure rate data is available for the device in question, and is implied at 90% statistical confidence (see 3.10). Also the "architecture" rules are modified as follows:

In addition, the following redundancy rules (7.4.4.3.1 of Part 2) will apply:

SIL 4 – Hardware Fault Tolerance of 2 (i.e. 1 out of 3, 2 out of 4 etc.)
SIL 3 – Hardware Fault Tolerance of 1 (i.e. 1 out of 2, 2 out of 3 etc.)
SIL 2 – Hardware Fault Tolerance of 0 (i.e. simplex but low demand only)
SIL 1 – Hardware Fault Tolerance of 0 (i.e. simplex low or high demand)

However, the majority of so called data tends to be based on manufacturers' warranty statistics or even FMEAs and does NOT qualify as field data. The authors therefore believe that invoking this rule is unlikely to be frequently justified.

3.3.3 Random Hardware Failures

Section 7.4.5

This is traditionally known as "reliability prediction" which, in the past, has dominated risk assessment work. It involves specifying the reliability model, the failure rates to be assumed, the component down times, diagnostic intervals and coverage. It is, of course, only a part of the picture since systematic failures must be addressed qualitatively via the rigor of life-cycle activities.

Techniques such as FMEA (failure mode and effect analysis), reliability block diagrams and fault tree analysis are involved and Chapters 5 and 6 together with Appendix 4 briefly describe how to carry these out. The Standard refers to confidence levels in respect of failure rates and this will be dealt with later.

In Chapter 1 we mentioned the anomaly concerning the allocation of the quantitative failure probability target to the random hardware failures alone. There is yet another anomaly concerning judgement of whether the target is met. If the fully quantified approach (described in Chapter 2) has been adopted then the failure target will be a PFD (probability of failure on demand) or a failure rate. The reliability prediction might suggest that the target is not met although still remaining within the PFD/rate limits of the SIL in question. The rule here is that since we have chosen to adopt a fully quantitative approach we should meet the target set (paragraph 7.4.5.1 of Part 2 of the Standard confirms this view). For example a PFD of 2×10^{-3} might have been targeted for a safety-related risk reduction system. This is of, course, SIL 2. The assessment might suggest that it will achieve 5×10^{-3} which is indeed SIL 2. However, since a target of 2×10^{-3} is the case then that target has NOT been met.

The question might then be asked "What if we had opted for a simpler risk graph approach and stated the requirement merely as a SIL – then would we not have met the requirement?" Indeed we have and this appears to be inconsistent. Once again there is no right or wrong answer to the dilemma. The Standard does not address it and, as in all such matters, the judgement of the responsible engineer is needed. Both approaches are admissible and, in any case, the accuracy of quantification is not very high (see Chapter 5).

3.4 Integration and Test (Referred to as Verification)

Section 7.5 and 7.9 of the Standard Table B3 [avoidance]

Based on the intended functionality the system should be tested, and the results recorded, to ensure that it fully meets the requirements. This is the type of testing which, for example, looks at the output responses to various combinations of inputs. This applies to all SILs.

Furthermore, a degree of additional testing, such as the response to unusual and "not specified" input conditions should be carried out. For SIL 1 and SIL 2 this should include system partitioning testing and boundary value testing. For SIL 3 and SIL 4 the tests should be extended to include test cases that combine critical logic requirements at operation boundaries.

3.5 Operations and Maintenance

Section 7.6 Table B4 [Avoidance]

(a) The system should have clear and concise operating and maintenance procedures. These procedures, and the safety system interface with personnel, should be designed to be user, and maintenance, friendly. This applies to all SIL levels.

(b) Documentation needs to be kept, of audits and for any proof-testing that is called for. There need to be records of the demand rate of the safety-related equipment, and furthermore failures also need to be recorded. These records should be periodically reviewed, to verify that the target safety integrity level was indeed appropriate and that it has been achieved. This applies to all SILs.

(c) For SIL 1 and SIL 2 systems, the operator input commands should be protected by key switches/passwords and all personnel should receive basic training. In addition, for SIL 3 and SIL 4 systems operating/maintenance procedures should be highly robust and personnel should have a high degree of experience and undertake annual training. This should include a study of the relationship between the safety-related system and the EUC.

3.6 Validation (Meaning Overall Acceptance Test and the Close Out-of Actions)

Section 7.3 and 7.7: Table B5

The object is to ensure that all the requirements of the safety system have been met and that all the procedures have been followed (albeit this should follow as a result of a company's functional safety capability).

A validation plan is needed which cross-references all the functional safety requirements to the various calculations, reviews and tests which verify the individual features. The completed cross-referencing of the results/reports provides the verification report. A spreadsheet is often effective for this purpose.

It is also necessary to ensure that any remedial action or additional testing arising from earlier tests has been carried out. In other words there is:

- A description of the problem (symptoms)
- A description of the causes
- The solution
- Evidence of re-testing to clear the problem.

This requirement applies to all SIL levels.

3.7 Safety Manuals

Section 7.4.9.3–7 and App D

For specific hardware or software items a safety manual is called for. Thus, instrumentation, PLCs and field devices will each need to be marketed with a safety manual. Re-useable items of code and software packages will also require a safety manual. Contents should include, for hardware (software is dealt with in the next chapter):

- a detailed specification of the functions
- the hardware and/or software configuration
- failure modes of the item
- for every failure mode an estimated failure rate
- failure modes that are detected by internal diagnostics
- failure modes of the diagnostics
- the hardware fault tolerance
- proof test intervals (if relevant).

3.8 Modifications

Section 7.8

For all modifications and changes there should be:

- revision control
- a record of the reason for the design change
- an impact analysis
- re-testing of the changed and any other affected modules.

The methods and procedures should be exactly the same as those applied at the original design phase. This paragraph applies to all SILs.

The Standard requires that, for SIL 1, changed modules are re-verified, for SIL 2 all affected modules are re-verified. For software (see Chapter 4) at SIL 3 the whole system is re-validated.

3.9 Acquired Sub-systems

For any sub-system which is to be used as part of the safety system, and is acquired as a complete item by the integrator of the safety system, there will need to be established, in addition to any other engineering considerations, the following parameters.

- Random hardware failure rates, categorized as:
  - fail safe failures
  - dangerous failures detected by auto-test
  - dangerous failures detected by proof test
- Procedures/methods for adequate proof testing
- The hardware fault tolerance of the sub-system
- The highest SIL that can be claimed as a consequence of the measures and procedures used during the design and implementation of the hardware and software or,
- A SIL derived by claim of "proven in use" see Paragraph 3.10 below.

3.10 "Proven in Use" (Referred to as Route 2_s in the Standard)

The Standard calls the use of the systematic techniques described in this chapter **route 1_s.** Proven-in-use is referred to as route 2_s. It also refers to route 3_s but this is, in fact a matter for Part 3.

As an alternative to all the systematic requirements summarized in this Chapter, adequate statistical data from field use may be used to satisfy the Standard. The random hardware failures prediction and safe failure fraction demonstrations are, however, still required. The previous field experience should be in an application and environment, which is very similar to the intended use. All failures experienced, whether due to hardware failures or systematic faults, should be recorded, along with total running hours. The Standard asks that the calculated failure rates should be claimed using a confidence limit of at least 70% (note that the 1_H rule asks for 90%).

Paragraph 7.4.10 of Part 2 allow for statistical demonstration that a SIL has been met in use. In Part 7 Annex D there are a number of pieces of statistical theory which purport to be appropriate to establishing confidence for software failures. However, the same theory applies to hardware failures and for the purposes of the single-sided 70% requirement can be summarized as follows.

For zero failures, the following "number of operations/demands" or "equipment hours" are necessary to infer that the lower limit of each SIL has been exceeded. Note that the operations and years should be field experience and not test hours or test demands.

SIL 1 (1: 10^{-1} or 10^{-1} per annum)	12 operations	or 12 years
SIL 2 (1: 10^{-2} or 10^{-2} per annum)	120 operations	or 120 years
SIL 3 (1: 10^{-3} or 10^{-3} per annum)	1200 operations	or 1200 years
SIL 4 (1: 10^{-4} or 10^{-4} per annum)	12000 operations	or 12000 years

For one failure, the following table applies. The times for larger numbers of failures can be calculated accordingly (i.e. from chi square methods).

SIL 1 (1: 10^{-1} or 10^{-1} per annum)	24 operations	or 24 years
SIL 2 (1: 10^{-2} or 10^{-2} per annum)	240 operations	or 240 years
SIL 3 (1: 10^{-3} or 10^{-3} per annum)	2400 operations	or 2400 years
SIL 4 (1: 10^{-4} or 10^{-4} per annum)	24000 operations	or 24000 years

The 90% confidence requirement would approximately double the experience requirement. The theory is dealt with in Smith DJ, Reliability, Maintainability and Risk.

3.11 ASICs and CPU Chips

(a) Digital ASICS and User Programmable ICs

Section 7.4.6.7 and Annex F of the Standard

All design activities are to be documented and all tools, libraries and production procedures should be proven in use. In the case of common or widely used tools, information about possible bugs and restrictions is required.

All activities and their results should be verified, for example by simulation, equivalence checks, timing analysis or checking the technology constraints.

For third party soft-cores and hard-cores, only validated macro blocks should be used and these should comply with all constraints and proceedings defined by the macro core provider if practicable. Unless already proven in use, each macro block should be treated as newly written code, for example it should be fully validated.

For the design, a problem-oriented and abstract high-level design methodology and design description language should be used. There should be adequate testability (for production test). Gate and interconnection (wire) delays should be considered.

Internal gates with tristate outputs should be avoided. If internal tristate outputs are used these outputs should be equipped with pull-ups/downs or bus-holders.

Before production, an adequate verification of the complete ASIC (i.e. including each verification step carried out during design and implementation to ensure correct module and chip functionality) should be carried out.

There are two tables in Annex F to cover Digital ASICs and Programmable ICs. They are very similar and are briefly summarized in one of the tables at the end of this chapter.

(b) Digital ICs With On-chip Redundancy (up to SIL 3)

Annex E of the Standard

A single IC semi-conductor substrate may contain on-chip redundancy subject to conservative constraints and given that there is a Safety Manual.

Establish separate physical blocks on the substratum of the IC for each channel and each monitoring element such as a watchdog. The blocks shall include bond wires and pin-out. Each channel shall have its own separated inputs and outputs which shall not be routed through another channel/block.

Take appropriate measures to avoid dangerous failure caused by faults of the power supply including common cause failures.

The minimum distance between boundaries of different physical blocks shall be sufficient to avoid short circuit and cross talk.

The substratum shall be connected to ground independent from the IC design process used (n-well or p-well).

The detection of a fault (by diagnostic tests, proof tests) in an IC with on-chip redundancy shall result in a safe state.

The minimum diagnostic coverage of each channel shall be at least 60%.

If it is necessary to implement a watchdog, for example for program sequence monitoring and/or to guarantee the required diagnostic coverage or safe failure fraction one channel shall not be used as a watchdog of another channel, except the use of functional diverse channels.

When testing for electromagnetic compatibility without additional safety margins the function carried out by the IC shall not be interfered with.

Avoid unsymmetrical wiring.

Beware of circuit faults leading to over-temperature.

For SIL 3 there shall be documented evidence that all application specific environmental conditions are in accordance with that taken into account during specification, analysis, verification and validation shall be provided. External measures are required that can achieve or maintain a safe state of the E/E/PE system. These measures require medium effectiveness as a minimum. All measures implemented inside the IC to monitor for effects of systematic and/or common cause failures shall use these external measures to achieve or maintain a safe state.

The Standard provides a CCF (Partial Beta type) Model. Partial Beta modeling is dealt with in Chapter 5.2.2. A Beta of 33% is taken as the starting point. Numbers are added or subtracted from this according to features which either compromise or defend against CCF. It is necessary to achieve a Beta of no greater than 25%. The scoring is provided in Appendix E of the Standard and summarized in the last table at the end of this chapter.

3.12 Conformance Demonstration Template

In order to justify that the requirements have been satisfied, it is necessary to provide a documented demonstration.

The following Conformance Demonstration Template is suggested as a possible format. The authors (as do many guidance documents) counsel against SIL 4 targets. In the event of such a case more rigorous detail from the Standard would need to be addressed.

IEC 61508 PART 2

For embedded software designs, with new hardware design, the demonstration might involve a reprint of all the tables from the Standard. The evidence for each item would then be entered in the right hand column as in the simple tables below.

However, the following tables might be considered adequate for relatively straightforward designs.

Under "Evidence" enter a reference to the project document (e.g. spec, test report, review, calculation) which satisfies that requirement. Under "Feature" take the text in conjunction with the fuller text in this chapter and/or the text in the IEC 61508 Standard. Note that a "Not Applicable" entry is acceptable if it can be justified.

The majority of the tables address "Procedures during the life-cycle". Towards the end there are tables which summarize "Techniques during the life-cycle".

General/life-cycle (Paras 7.1, 7.3) (Table '1')

Feature (all SILs)	**Evidence**
Existence of a Quality and Safety Plan (see Appendix 1), including document hierarchy, roles and competency, validation plan etc	
Description of overall novelty, complexity, reason for SIL targets, rigor needed etc	
Clear documentation hierarchy (Q&S Plan, Functional Spec, Design docs, Review strategy, Integration and Test plans etc)	
Adequately cross-referenced documents which identify the FS requirements.	
Adequate project management as per company's FSM procedure	
The project plan should include adequate plans to validate the overall requirements. It should state the state tools and techniques to be used.	
Feature (SIL 3)	
Enhanced rigor of project management and appropriate independence	

Specification (Para 7.2) (Table B1)

Feature (all SILs)	**Evidence**
Clear text and some graphics, use of checklist or structured method, precise, unambiguous. Describes SR functions and separation of EUC/SRS, responses, performance requirements, well defined interfaces, modes of operation.	
SIL for each SR function, high/low demand	
Emc addressed	
Either: Inspection of the spec, semi-formal methods, checklists, CAS tool or formal method	
Feature (SIL 2 and above)	**Evidence**
Inspection/review of the specification	
Feature (SIL 3)	**Evidence**
Use of a semi-formal method	
Physical separation of EUC/SRS	

Design and development (Para 7.4) (Tables B2, A15–A18)

Feature (all SILs)	Evidence
Use of in-house design standards and work instructions	
Sector specific guidance addressed as required	
Visible and adequate design documentation	
Structured design in evidence	
Proven components and subsystems (justified by 10 for 1 year)	
Modular approach with SR elements independent of nonSR and interfaces well defined.	
SR SIL = Highest of mode SILs	
Adequate component de-rating (in-house or other standards)	
Non-SR failures independent of SRS	
Safe state achieved on detection of failure	
Data-communications errors addressed	
No access by user to change hardware or software	
Operator interfaces considered	
Fault tolerant technique (minimum of a watchdog)	
Appropriate emc measures	
Feature (SIL 2 and above)	**Evidence**
Checklist or walkthrough or design tools	
Higher degree of fault tolerance	
Appropriate emc measures as per Table A17	
Feature (SIL 3)	**Evidence**
Use of semi-formal methods	
Proven components and subsystems (certified or justified by 10 for 2 year)	
Higher degree of fault tolerance and monitoring (e.g. memory checks)	

Random hardware failures and architectures (Paras 7.4.4, 7.4.5)

Feature (all SILs)	Evidence
SFF and architectural conformance is to be demonstrated OR alternative route (proven-in-use)	
Random hardware failures are to be predicted and compared with the SIL or other quantified target	

Random hardware failures assessment contains all the items suggested in Appendix 2 of this book. Include Reliability model, CCF model, justification of choice of failure rate data, coverage of all the hazardous failure modes	
Feature (SFF = > 90%)	**Evidence**
SFF assessed by a documented FMEA (adequate rigor Appendix 4)	
Appropriate choice of Type A or Type B SFF Table	
Feature (SIL 3)	
Fault insertion (sample) in the FMEA process	

Integration and test (Paras 7.5, 7.9) (Table B3)

Feature (all SILs)	**Evidence**
Overall review and test strategy in Q&S Plan	
Test specs, logs of results and discrepancies, records of versions, acceptance criteria, tools	
Evidence of remedial action	
Functional test including input partitioning, boundary values, unintended functions and non-specified input values	
Feature (SIL 2 and above)	**Evidence**
As for SIL 1	
Feature (SIL 3)	**Evidence**
Include tests of critical logic functions at operational boundaries	
Standardized procedures	

Operations and maintenance (Para 7.6) (Table B4)

Feature (all SILs)	**Evidence**
Safety Manual in place - if applicable	
Component wear out life accounted for by preventive replacement Proof tests specified	
Procedures validated by Ops and Mtce staff Commissioning successful	

(Continued)

Feature (all SILs)	**Evidence**
Failures (and Actual Demands) reporting procedures in place	
Start-up, shut-down and fault scenarios covered	
User friendly interfaces	
Lockable switch or password access	
Operator i/ps to be acknowledged	
Basic training specified	
Feature (SIL 2 and above)	**Evidence**
Protect against operator errors OR specify operator skill	
Feature (SIL 3)	**Evidence**
Protect against operator errors AND specify operator skill	
At least annual training	

Validation (Para 7.7) (Table B5)

Feature (all SILs)	**Evidence**
Validation plan actually implemented. To include:	
Function test	
Environmental test	
Fault insertion	
Calibration of equipment	
Records and close-out report	
Discrepancies positively handled	
Functional tests	
Environmental tests	
Interference tests	
Fault insertion	
Feature (SIL 2 and above)	**Evidence**
Check all SR functions OK in presence of faulty operating conditions	
Feature (SIL 3)	**Evidence**
Fault insertion at unit level	
Some static or dynamic analysis or simulation	

Modifications (Para 7.8)

Feature (all SILs)	Evidence
Change control with adequate competence	
Impact analysis carried out	
Re-verify changed modules	
Feature (SIL 2 and above)	**Evidence**
Re-verify affected modules	

Acquired sub-systems

Feature (at the appropriate SIL)	Evidence
SIL requirements reflected onto suppliers	
Compliance demonstrated	

Proven in use (Para 7.10)

Feature (at the appropriate SIL)	Evidence
Application appropriate and restricted functionality	
Any differences to application addressed and conformance demonstrated	
Statistical data available at 70% confidence to verify random hardware failures target	
Failure data validated	

Techniques (ASICs & ICs) (Annexe F) (Summary)

In general, the following summary can be assumed to apply for all SILs. The Standard provides some graduation in the degrees of effectiveness.

Design phase	Technique/measure	Evidence
Design entry	Structured description in (V)HDL* with proven simulators	
	Functional test on module and top level	
	Restricted use of asynchronous constructs	
	Synchronization of primary inputs and control of metastability	
	Coding guidelines with defensive programming	
	Modularization	

(Continued)

Design phase	Technique/measure	Evidence
	Design for testability	
	Use of Boolean if programmable ICs	
Synthesis	Simulation of the gate netlist, to check timing constraints or	
	Static analysis of the propagation delay (STA)	
	Internal consistency checks	
	Verification of the gate netlist	
	Application of proven in use synthesis tools & libraries	
Test insertion and test pattern generation	Implementation of test structures and estimation of the test coverage by simulation (ATPG tool)	
	Simulation of the gate netlist, to check timing constraints or verification against reference model	
Placement, routing, layout generation	Proven in use or validated hard cores with online testing	
	Simulation or verification of the gate netlist, to check timing constraints or Static analysis of the propagation delay (STA)	
Chip production	Proven in use process technology with QA	

*Very high speed integrated circuit hardware description

Assessment of CCF (CPUs) (Annex E) see 3.11

Technique/measure decreasing β	β -actor [%]
Diverse measures or functions in different channels	4–6
Testing for emc with additional safety margin)	5
Providing each block with its own power supply pins - no block supplied via another	6
Isolate and decouple physical locations	2–4
Ground pin between pin-out of different blocks	2
High diagnostic coverage (≥ 99%) of each channel	7–9
Technique/measure increasing β	β-factor [%]
Watchdog on-chip used as monitoring element	5
Monitoring elements on-chip other than watchdog, for example clock monitoring	5–10
Internal connections between blocks by wiring between output and input cells of different blocks without cross-over	2
Internal connections between blocks by wiring between output and input cells of different blocks with cross-over	4

CHAPTER 4

Meeting IEC 61508 Part 3

Chapter Outline

Safety Critical Systems Handbook. DOI: 10.1016/B978-0-08-096781-3.10004-5

IEC 61508 Part 3 covers the development of software. This chapter summarizes the main requirements. However, the following points should be noted first.

Whereas the reliability prediction of hardware failures, addressed in Section 3.3.3 of the last chapter, predicts a failure rate to be anticipated, the application and demonstration of qualitative measures DOES NOT imply a failure rate for the systematic failures. All that can be reasonably claimed is that, given the state of the art, we believe the measures specified are appropriate for the integrity level in question and that therefore the systematic failures will credibly be similar to and not exceed the hardware failure rate of that SIL.

The Annexes of Part 3 offer appropriate techniques, by SIL, in the form of tables followed by more detailed tables with cross-references. In the 2010 version there is an additional Annex giving guidance on the properties that the software techniques should achieve which is intended to allow a frame work for justifying alternative techniques to those given in the standard.

This chapter attempts to provide a simple and useable interpretation. At the end of this chapter a "conformance demonstration template" is suggested which, when completed for a specific product or system assessment, will offer evidence of conformance to the SIL in question

The approach to the assessment will differ substantially between:

EMBEDDED SOFTWARE DESIGN

and

APPLICATIONS SOFTWARE

The demonstration template tables at the end of this chapter cater for the latter case. Chapter 8, which will cover the restricted subset of IEC 61511, also caters for applications software.

4.1 Organizing and Managing the Software Engineering

4.1.1 Section 7.1 and Annex G of the Standard Table '1'

Section 3.1 of the previous chapter applies here in exactly the same way and therefore we do not repeat it.

In addition, the Standard recommends the use of the 'V' model approach to software design, with the number of phases in the 'V' model being adapted according to the target safety integrity level and the complexity of the project. The principle of the 'V' model is a top-down design approach starting with the 'overall software safety specification' and ending, at the bottom, with the actual software code. Progressive testing of the system starts with the lowest level of software module, followed by integrating modules, and working up to testing the complete safety system. Normally, a level of testing for each level of design would be required.

The life-cycle should be described in writing (and backed up by graphical figures such as are shown in Figures 4.1– 4.3). System and hardware interfaces should be addressed and it should reflect the architectural design. The 'V' model is frequently quoted and is illustrated in Figure 4.1. However, this is somewhat simplistic and Figures 4.2 and 4.3 show typical interpretations of this model as they might apply to the two types of development mentioned in the box at the beginning of this chapter. Beneath each of the figures is a statement describing how they meet the activities specified in the Standard.

Figure 4.2 describes a simple proven PLC platform with ladder logic code providing an application such as process control or shut down. Figure 4.3 describes a more complex development where the software has been developed in a high-level language (for example a C subset or Ada) and where there is an element of assembler code.

Other life-cycle models, like the 'Waterfall', are acceptable provided they incorporate the same type of properties as the V model. At SIL 2 and above there needs to be evidence of positive justifications and reviews of departures from the life-cycle activities listed in the Standard.

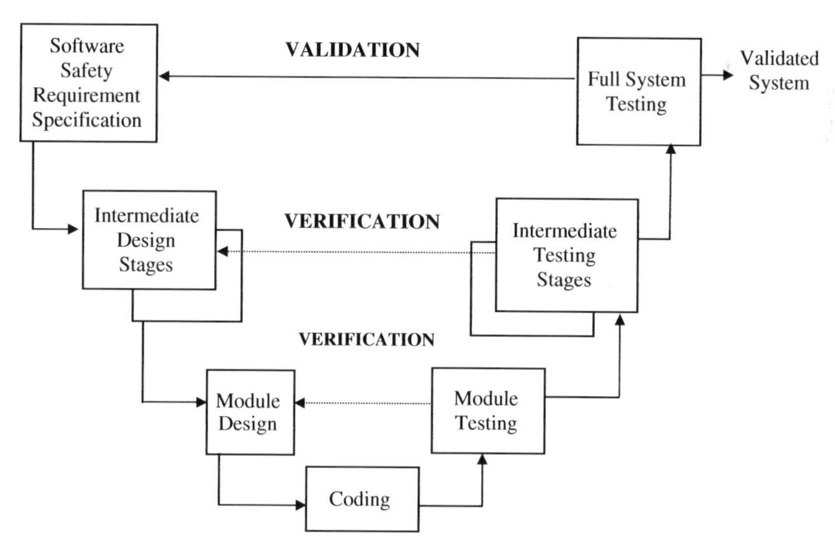

Figure 4.1: A typical 'V' model.

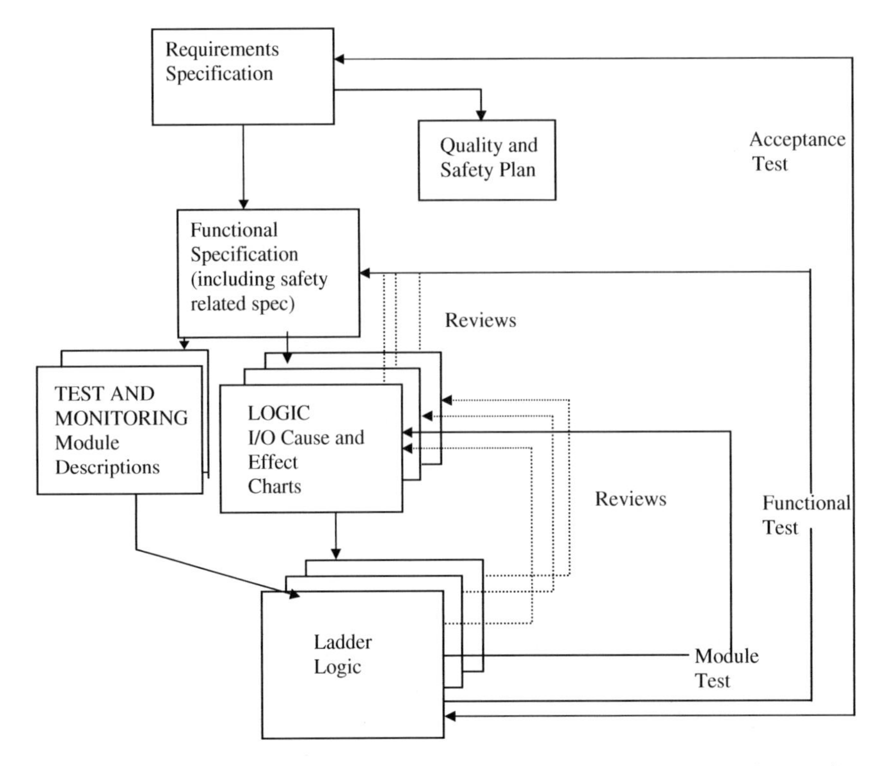

Figure 4.2: A software development life-cycle for a simple PLC system at the application level. The above life-cycle model addresses the architectural design in the Functional Specification and the module design by virtue of cause and effect charts. Integration is a part of the functional test and validation is achieved by means of acceptance test and other activities listed in the Quality and Safety Plan.

Annex G provides guidance on tailoring the life-cycle for "data driven systems". Some systems are designed in two parts:

- A basic system with operating functions
- A data part which defines/imposes an application onto the basic system.

The amount of rigor needed will depend on the complexity of the behavior called for by the design. This complexity can be classified into classes as follows:

- Variability allowed by the language:
  - fixed program
  - limited variability
  - full variability
- Ability to configure application:
  - limited
  - full.

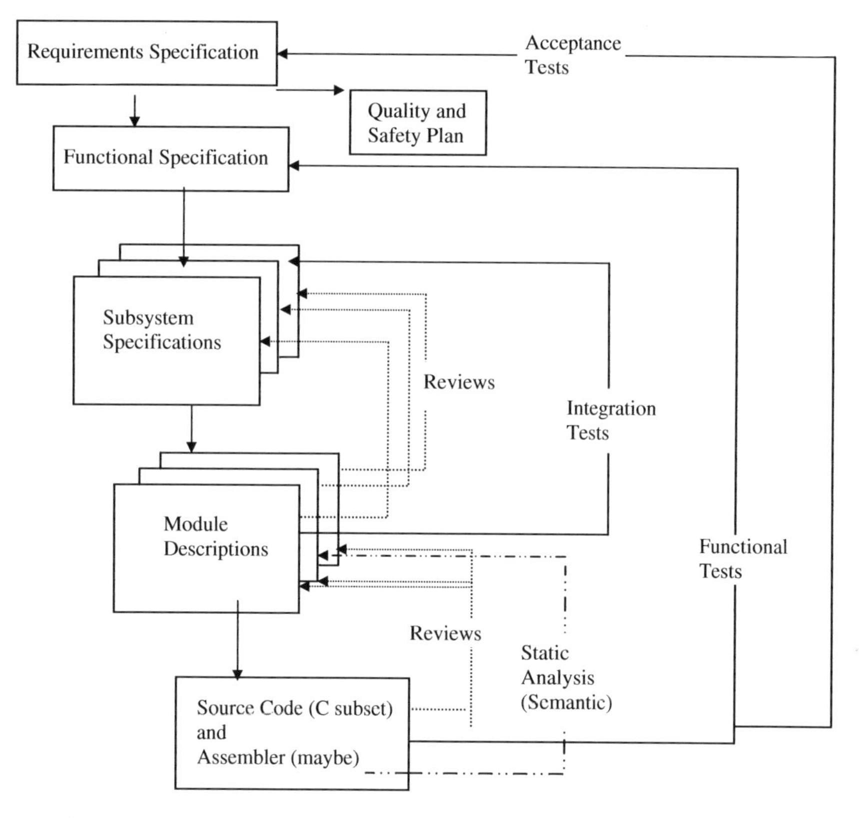

Figure 4.3: A software development lifecycle for a system with embedded software. The above life-cycle model addresses the architectural design in the Functional Specification. Validation is achieved by means of acceptance test and other activities listed in the Quality and Safety Plan.

A brief summary of these is provided in Annex G and is summarized at the end of this chapter.

The software configuration management process needs to be clear and to specify:

- Levels where configuration control commences
- Where baselines will be defined and how they will be established
- Methods of traceability of requirements
- Change control
- Impact assessment
- Rules for release and disposal.

At SIL 2 and above configuration control must apply to the smallest compiled module or unit.

4.2 Requirements Involving the Specification

Section 7.2 of the Standard: Table A1

(a) The software safety requirements, in terms of both the safety functions and the safety integrity, should be stated in the software safety requirements specification. Items to be covered include:

- Capacities and response times
- Equipment and operator interfaces including misuse
- Software self monitoring
- Functions which force a safe state
- Overflow and underflow of data storage
- Corruption
- Out of range values
- Periodic testing of safety functions whilst system is running.

(b) The specification should **include** all the modes of operation, the capacity and response time performance requirements, maintenance and operator requirements, self monitoring of the software and hardware as appropriate, enabling the safety function to be testable whilst the EUC is operational, and details of all internal/external interfaces. The specification should extend down to the configuration control level.

(c) The specification should be written in **a clear and precise** manner, traceable back to the safety specification and other relevant documents. The document should be free from ambiguity and clear to those for whom it is intended.

For SIL 1 and SIL 2 systems, this specification should use semi-formal methods to describe the critical parts of the requirement (e.g. safety-related control logic). For SIL 3 and SIL 4, semi-formal methods should be used for all the requirements and, in addition, at SIL 4 there should be the use of computer support tools for the critical parts (e.g. safety-related control logic).

Forwards and backwards traceability should be addressed.

The semi-formal methods chosen should be appropriate to the application and typically include logic/function block diagrams, cause and effect charts, sequence diagrams, state transition diagrams, time Petri nets, truth tables and data flow diagrams.

4.3 Requirements for Design and Development

4.3.1 Features of the Design and Architecture

Section 7.4.3 of the Standard: Table A2

(a) The design methods should aid modularity and embrace features which reduce complexity and provide clear expression of functionality, information flow, data structures, sequencing, timing related constraints/information, and design assumptions.

(b) The system software (i.e. non-application software) should include software for diagnosing faults in the system hardware, error detection for communication links, and on-line testing of standard application software modules.

In the event of detecting an error or fault the system should, if appropriate, be allowed to continue but with the faulty redundant element or complete part of the system isolated.

For SIL 1 and SIL 2 systems there should be basic hardware fault checks (i.e. watchdog and serial communication error detection).

For SIL 3 and SIL 4, there needs to be some hardware fault detection on all parts of the system, i.e. sensors, input/output circuits, logic resolver, output elements and both the communication and memory should have error detection.

(c) Non-interference (i.e. where a system hosts both non-safety-related and a safety-related functions) then Annex F provides a list of considerations such as:

- shared use of RAM, peripherals & processor time
- communications between elements
- failures in an element causing consequent failure in another.

4.3.2 Detailed Design and Coding

Paragraphs 7.4.5, 7.4.6, Tables A4, B1, B5, B7, B9

(a) The detailed design of the software modules and coding implementation should result in small manageable software modules. Semi-formal methods should be applied, together with design and coding standards including structured programming, suitable for the application. This applies to all SILs.

(b) The system should, as far as possible, use trusted and verified software modules, which have been used in similar applications. This is called for from SIL 2 upwards.

(c) The software should not use dynamic objects, which depend on the state of the system at the moment of allocation, where they do not allow for checking by offline tools. This applies to all SILs.

(d) For SIL 3 and SIL 4 systems, the software should include additional defensive programming (e.g. variables should be both range and, where possible, plausibility checked). There should also be limited use of interrupts, pointers, and recursion.

4.3.3 Programming Language and Support Tools

Paragraph 7.4.4, Table A3

(a) The programming language should be capable of being fully and unambiguously defined. The language should be used with a specific coding standard and a restricted sub-set, to minimize unsafe/unstructured use of the language. This applies to all SILs.

At SIL 2 and above, dynamic objects and unconditional branches should be forbidden. At SIL 3 and SIL 4 more rigorous rules should be considered such as the limiting of interrupts and pointers, and the use of diverse functions to protect against errors which might arise from tools.

(b) The support tools need either to be well proven in use (and errors resolved) and/or certified as suitable for safety system application. The above applies to all SILs, with certified tools more strongly recommended for SIL 3 and SIL 4.

(c) The requirements for support tools should also apply to off-line software packages that are used in association with any design activity during the safety life cycle. An example of this would be a software package that is used to perform the safety loop PFD or failure rate calculation. These tools need to have been assessed to confirm both completeness and accuracy and there should be a clear instruction manual.

4.4 Integration and Test (Referred to as Verification)

4.4.1 Software Module Testing and Integration

Paragraphs 7.4.7, 7.4.8, Tables A5, B2, B3, B6, B8

(a) The individual software modules should be code reviewed and tested to ensure that they perform the intended function and, by a selection of limited test data, to confirm that the system does not perform unintended functions.

(b) As the module testing is completed then module integration testing should be performed with pre-defined test cases and test data. This testing should include functional, "black box", and performance testing.

(c) The results of the testing should be documented in a chronological log and any necessary corrective action specified. Version numbers of modules and of test instructions should be clearly indicated. Discrepancies from the anticipated results should be clearly visible. Any modifications or changes to the software which are implemented after any phase of the testing should be analysed to determine the full extent of re-test that is required.

(d) The above needs to be carried out for all SILs; however, the extent of the testing for unexpected and fault conditions needs to be increased for the higher SILs. As an example, for

SIL 1 and SIL 2 systems the testing should include boundary value testing and partitioning testing and in addition, for SIL 3 and SIL 4, tests generated from cause consequence analysis of certain critical events.

4.4.2 Overall Integration Testing

Paragraph 7.5, Table A6

These recommendations are for testing the integrated system, which includes both hardware and software and, although this requirement is repeated in Part 3, the same requirements have already been dealt with in Part 2.

This phase continues through to Factory Acceptance Test. Test harnesses are part of the test equipment and require adequate design documentation and proving. Test records are vital as they are the only visibility to the results.

4.5 Validation (Meaning Overall Acceptance Test and Close Out of Actions)

Paragraphs 7.3, 7.7, 7.9, Table A79

(a) Whereas verification implies confirming, for each stage of the design, that all the requirements have been met prior to the start of testing of the next stage (shown in Figures 4.2–4.3), validation is the final confirmation that the total system meets all the required objectives and that all the design procedures have been followed. The Functional Safety Management requirements (Chapter 2) should cover the requirements for both validation and verification.

(b) The Validation plan should show how all the safety requirements have been fully addressed. It should cover the entire life-cycle activities and will show audit points. It should address specific pass/fail criteria, a positive choice of validation methods and a clear handling of non-conformances.

(c) At SIL 2 and above some test coverage metric should be visible. At **SILs 3 and 4** a more rigorous coverage of accuracy, consistency, conformance with standards (e.g. coding rules) is needed.

4.6 Safety Manuals

(Annex D)

For specific software elements which are re-used, a safety manual is called for. Its contents shall include:

- A description of the element and its attributes
- Its configuration and all assumptions

- The minimum degree of knowledge expected of the integrator
- Degree of reliance placed on the element
- Installation instructions
- The reason for release of the element
- Details of whether the pre-existing element has been subject to release to clear outstanding anomalies, or inclusion of additional functionality
- Outstanding anomalies
- Backward compatibility
- Compatibility with other systems
- A pre-existing element may be dependent upon a specially developed operating system
- The build standard should also be specified incorporating compiler identification and version, tools
- Details of the pre-existing element name(s) and description(s) should be given, including the version / issue / modification state
- Change control
- The mechanism by which the integrator can initiate a change request
- Interface constraints
- Details of any specific constraints, in particular user interface requirements shall be identified
- A justification of the element safety manual claims.

4.7 Modifications

Paragraph 7.6, 7.8, Table A 8 and B9

(a) The following are required:

- A modification log
- Revision control
- Record of the reason for design change
- Impact analysis
- Re-testing as in (b) below.

The methods and procedures should be at least equal to those applied at the original design phase. This paragraph applies for all SIL levels.

The modification records should make it clear which documents have been changed and the nature of the change.

(b) For SIL 1 changed modules are re-verified, for **SIL 2** all affected modules are re-verified and for **SIL 3 and above** the whole system needs to be re-validated. This is not trivial and may add considerably to the cost for a SIL 3 system involving software.

4.8 Alternative Techniques and Procedures

Annex C of the 2010 version provides guidance on justifying the properties that alternative software techniques should achieve. The properties to be examined, in respect of a proposed alternative technique, are:

- Completeness with respect to the safety needs
- Correctness with respect to the safety needs
- Freedom from specification faults or ambiguity
- Ease by which the safety requirements can be understood
- Freedom from adverse interference from non-safety software
- Capability of providing a basis for verification and validation.

The methods of assessment (listed in Annex C) are labeled R1, R2, R3 and "-".

- For SIL1/2: R1 – limited objective acceptance criteria (e.g. black box test, field trial)
- For SIL3: R2 – objective acceptance criteria with good confidence (e.g. tests with coverage metrics)
- For SIL4: R3 - objective systematic reasoning (e.g. formal proof)
- "-" not relevant.

4.9 Data Driven Systems

This is where the applications part of the software is written in the form of data which serves to configure the system requirements/functions. Annex G covers this as follows

4.9.1 Limited Variability Configuration, Limited Application Configurability

The configuration language does not allow the programmer to alter the function of the system but is limited to adjustment of data parameters (e.g. SMART sensors and actuators). The justification of the tailoring of the safety lifecycle should include the following:

(a) specification of the input parameters;
(b) verification that the parameters have been correctly implemented;
(c) validation of all combinations of input parameters;
(d) consideration of special and specific modes of operation during configuration;
(e) human factors/ergonomics;
(f) interlocks, e.g. ensuring that operational interlocks are not invalidated during configuration;
(g) inadvertent re-configuration, e.g. key switch access, protection devices.

4.9.2 Limited Variability Configuration, Full Application Configurability

As above but can create extensive static data parameters (e.g. an air traffic control system). In addition to the above the justification shall include:

(a) automation tools for creation of data;
(b) consistency checking, e.g. the data is self compatible;
(c) rules checking, e.g. to ensure the generation of data meets the constraints;
(d) validity of interfaces with the data preparation systems.

4.9.3 Limited Variability Programming, Limited Application Configurability

These languages allow the user limited flexibility to customize the functions of the system to their own specific requirements, based on a range of hardware and software elements (e.g. functional block programming, ladder logic, spreadsheet-based systems).

In addition to the above two paragraphs the following should be included:

(a) the specification of the application requirements;
(b) the permitted language sub-sets for this application;
(c) the design methods for combining the language sub-sets;
(d) the coverage criteria for verification addressing the combinations of potential system states.

4.9.4 Limited Variability Programming, Full Application Configurability

The essential difference from limited variability programming, limited application configurability is complexity (e.g. graphical systems and SCADA-based batch control systems). In addition to the above paragraphs, the following should be included:

(a) the architectural design of the application;
(b) the provision of templates;
(c) the verification of the individual templates;
(d) the verification and validation of the application.

4.10 Some Technical Comments

4.10.1 Static Analysis

Static analysis is a technique (usually automated) which does not involve execution of code but consists of algebraic examination of source code. It involves a succession of "procedures" whereby the paths through the code, the use of variables and the algebraic functions of the algorithms are analysed. There are packages available which carry out the procedures and, indeed, modern compilers frequently carry out some of the static analysis procedures such as data flow analysis.

Table B8 of Part 3 lists Data flow and Control flow as HR (highly recommended) for SIL 3 and SIL 4. It should be remembered, however, that static analysis packages are only available for procedural high-level languages and require a translator which is language specific. Thus, static analysis cannot be automatically applied to PLC code other than by means of manual code walkthrough, which loses the advantages of the 100% algebraic capability of an automated package.

Semantic analysis, whereby functional relationships between inputs and outputs are described for each path, is the most powerful of the static analysis procedures. It is, however, not trivial and might well involve several man-days of analysis effort for a 500-line segment of code. It is not referred to in the Standard.

Static analysis, although powerful, is not a panacea for code quality. It only reflects the functionality in order for the analyst to review the code against the specification. Furthermore it is concerned only with logic and cannot address timing features.

It is worth noting that, in Table B8, design review is treated as an element of static analysis. It is, in fact, a design review tool.

If it is intended to use static analysis then some thought must be given as to the language used for the design because static analysis tools are language specific.

4.10.2 Use of "Formal" Methods

Table B5 of Part 3 refers to formal methods and Table A9 to formal proof. In both cases it is HR (highly recommended) for SIL 4 and merely R (recommended) for SIL 2 and SIL 3.

The term Formal Methods is much used and much abused. In software engineering it covers a number of methodologies and techniques for specifying and designing systems, both non-programmable and programmable. These can be applied throughout the life-cycle including the specification stage and the software coding itself.

The term is often used to describe a range of mathematical notations and techniques applied to the rigorous definition of system requirements which can then be propagated into the subsequent design stages. The strength of formal methods is that they address the requirements at the beginning of the design cycle. One of the main benefits of this is that formalism applied at this early stage may lead to the prevention, or at least early detection, of incipient errors. The cost of errors revealed at this stage is dramatically less than if they are allowed to persist until commissioning or even field use. This is because the longer they remain undetected the potentially more serious and far-reaching are the changes required to correct them.

The potential benefits may be considerable but they cannot be realized without properly trained people and appropriate tools. Formal methods are not easy to use. As with all languages, it is easier to read a piece of specification than it is to write it. A further complication is the choice of method for a particular application. Unfortunately, there is not a universally suitable method for all situations.

4.10.3 PLCs (Programmable Logic Controllers) and their Languages

In the past, PLC programming languages were limited to simple code (e.g. Ladder Logic) which is a limited variability language usually having no branching statements. These earlier languages are suitable for use at all SILs with only minor restrictions on the instruction set.

Currently PLCs have wider instruction sets, involving branching instructions etc., and restrictions in the use of the language set are needed at the higher SILs.

With the advent of IEC 61131-3 there is a range of limited variability programming languages and the choice will be governed partly by the application. Again restricted subsets may be needed for safety related applications. Some application specific languages are now available, as for example, the facility to program plant shutdown systems directly by means of Cause and Effect Diagrams. Inherently, this is a restricted subset created for safety-related applications.

4.10.4 Software Re-use

Parts 2 and 3 of the Standard refer to "trusted/verified", "proven in use" and "field experience" in various tables and in parts of the text. They are used in slightly different contexts but basically refer to the same concept of empirical evidence from use. However, "trusted/verified" also refers to previously designed and tested software without regard for its previous application and use.

Table A4 of Part 3 lists the re-use of "trusted/verified" software modules as "highly recommended" for SIL 2 and above.

It is frequently assumed that the re-use of software, including specifications, algorithms and code, will, since the item is proven, lead to fewer failures than if the software were developed anew. There are reasons for and against this assumption.

Reasonable expectations of reliability, from re-use, are suggested because:

- The re-used code or specification is proven
- The item has been subject to more than average test
- The time saving can be used for more development or test
- The item has been tested in real applications environments
- If the item has been designed for re-use it will be more likely to have stand-alone features such as less coupling.

On the other hand:

- If the re-used item is being used in a different environment undiscovered faults may be revealed
- If the item has been designed for re-use it may contain facilities not required for a particular application, therefore the item may not be ideal for the application and it may have to be modified
- Problems may arise from the internal operation of the item not being fully understood.

In Part 3, Paragraph 7.4.7.2 (Note 3) allow for statistical demonstration that a SIL has been met in use for a module of software. In Part 7 Annex D there are a number of pieces of statistical theory which purport to be appropriate to the confidence in software. However, the same statistical theory applies as with hardware failure data (Chapter 3.10).

In conclusion, provided that there is adequate control involving procedures to minimize the effects of the above then significant advantages can be gained by the re-use of software at all SILs.

4.10.5 Software Metrics

The term metrics, in this context, refers to measures of size, complexity and structure of code. An obvious example would be the number of branching statements (in other words a measure of complexity), which might be assumed to relate to error rate. There has been interest in this activity for many years but there are conflicting opinions as to its value.

The pre-2010 Standard mentions software metrics but merely lists them as "recommended" at all SILs. In the long term metrics, if collected extensively within a specific industry group or product application, might permit some correlation with field failure performance and safety-integrity. It is felt, however, that it is still "early days" in this respect.

The term metrics is also used to refer to statistics about test coverage, as called for in earlier paragraphs.

4.11 Conformance Demonstration Template

In order to justify that the requirements have been satisfied, it is necessary to provide a documented demonstration.

The following Conformance Demonstration Template is suggested as a possible format, addressing up to SIL 3 applications. The authors (as do many guidance documents) counsel against SIL 4 targets. In the event of such a case more rigorous detail from the Standard would need to be addressed.

IEC 61508 PART 3

For embedded software designs, with new hardware design, the demonstration might involve a reprint of all the tables from the Standard. The evidence for each item would then be entered in the right hand column as in the simple tables below.

However, the following tables might be considered adequate for relatively straightforward designs.

Under "Evidence" enter a reference to the project document (e.g. spec, test report, review, calculation) which satisfies that requirement. Under "Feature" take the text in conjunction with the fuller text in this chapter and/or the text in the IEC 61508 Standard. Note that a "Not applicable" entry is acceptable if it can be justified.

General (Paras 7.1, 7.3) (Table '1')

Feature (all SILs)	**Evidence**
Existence of S/W development plan including:	
Procurement, development, integration, verification, validation and modification activities	
Rev number, config management, config items, deliverables, responsible persons	
Evidence of review	
Description of overall novelty, complexity, SILs, rigor needed etc	
Clear documentation hierarchy (Q&S Plan, Functional Spec, Design docs, Review strategy, Integration and test plans etc etc)	
Adequate configuration management as per company's FSM procedure	
Feature (SIL 3)	
Enhanced rigor of project management and appropriate independence	

Life-cycle (Paras 7.1, 7.3) (Table '1')

Feature (all SILs)	**Evidence**
A Functional Safety audit has given a reasonable indication that the life-cycle activities required by the company's FSM procedure have been implemented	
The project plan should include adequate plans to validate the overall requirements and state tools and techniques.	
Adequate software life-cycle model as per this chapter including the document hierarchy	
Configuration management (All Documents and Media) specifying baselines, minimum configuration stage, traceability, release etc	

Feature (SIL 2 and above)	
Alternative life-cycle models to be justified	
Configuration control to level of smallest compiled unit	
Feature (SIL 3)	
Alternative life-cycle models to be justified and at least as rigorous	
Sample review of configuration status	

Specification (Para 7.2) (Table A1) [Table B7 amplifies semi-formal methods]

Feature (all SILs)	**Evidence**
There is a software safety requirements specification including:	
Revision number, config control, author(s) as specified in the Q&S plan	
Reviewed, approved, derived from Func Spec	
All modes of operation considered, support for FS and nonFS functions clear	
External interfaces specified	
Baselines and change requests	
Clear text and some graphics, use of checklist or structured method, Complete, precise, unambiguous and traceable	
Describes SR functions and their separation, performance requirements, well defined interfaces, all modes of operation	
Requirements uniquely identified and traceable	
Capacities and response times declared	
Adequate self monitoring and self test features addressed to achieve the SFF required	
A review of the feasibility of requirements by the software developer	
Feature (SIL 2 and above)	**Evidence**
Inspection of the specification (traceability to interface specs)	
Either computer aided spec tool or semi-formal method	
Feature (SIL 3)	**Evidence**
Use of a semi-formal method or tool and appropriately used (i.e. systematic representation of the logic throughout the spec)	
Traceability between system safety requirements, software safety requirements and the perceived safety needs	

Architecture and fault tolerance (Para 7.4.3) (Table A2)

Feature (all SILs)	Evidence
Major elements of the software, and their interconnection (based on partitioning) well defined	
Modular approach and clear partitioning into functions	
Use of structured methods in describing the architecture	
Address graceful degradation (i.e. resilience to faults)	
Program sequence monitoring (i.e. a watchdog function)	
Feature (SIL 2 and above)	**Evidence**
Clear visibility of logic (i.e. the algorithms)	
Determining the software cycle behavior and timing	
Feature (SIL 3)	**Evidence**
Fault detection and diagnosis	
Program sequence monitoring (i.e. counters and memory checks)	
Use of a semi-formal method	
Static resource allocation and synchronization with shared resource	

Design and development (Paras 7.4.5, 7.4.6) (Tables A2, A4, B1, B9)

Feature (all SILs)	Evidence
Structured S/W design, recognized methods, under config management	
Use of standards and guidelines	
Visible and adequate design documentation	
Modular design with minimum complexity whose decomposition supports testing	
Readable, testable code (each module reviewed)	
Small manageable modules (and modules conform to the coding standards)	
Diagnostic software (e.g. watchdog and comms checks)	
Isolate and continue on detection of fault	
Structured methods	
Feature (SIL 2 and above)	**Evidence**
Trusted and verified modules	
No dynamic objects, limited interrupts, pointers and recursion	

(Continued)

Feature (all SILs)	Evidence
No unconditional jumps	
Feature (SIL 3)	**Evidence**
Computer aided spec tool	
Semi-formal method	
Graceful degradation	
Defensive programming (e.g. range checks)	
No (or online check) dynamic variables	
Limited pointers, interrupts, recursion	

Language and support tools (Para 7.5) Table A3

Feature (all SILs)	Evidence
Suitable strongly types language	
Language fully defined, seen to be error free, unambiguous features, facilitates detection of programming errors, describes unsafe programming features	
Coding standard/manual (fit for purpose and reviewed)	
Confidence in tools	
Feature (SIL 2 and above)	**Evidence**
Certified tools or proven in use to be error free	
Trusted module library	
No dynamic objects	
Feature (SIL 3)	
Language subset (e.g. limited interrupts and pointers)	

Integration and test (Paras 7.4.7, 7.4.8, 7.5) (Tables A5, A6, B2, B3)

Feature (all SILs)	Evidence
Overall test strategy in Q&S Plan showing steps to integration and including test environment, tools and provision for remedial action	
Test specs, reports/results and discrepancy records and remedial action evidence	
Test logs in chronological order with version referencing	
Module code review and test (documented)	

(*Continued*)

Feature (all SILs)	Evidence
Integration tests with specified test cases, data and pass/fail criteria	
Pre-defined test cases with boundary values	
Response times and memory constraints	
Functional and black box testing	
Feature (SIL 2 and above)	**Evidence**
Dynamic testing	
Unintended functions tested on critical paths and formal structured test management	
Feature (SIL 3)	**Evidence**
Performance and interface testing	
Avalanche/stress tests	

Operations and maintenance (Para 7.6) (Table B4)

Feature (all SILs)	Evidence
Safety Manual in place - if applicable	
Proof tests specified	
Procedures validated by Ops and Mtce staff	
Commissioning successful	
Failures (and Actual Demands) reporting procedures in place	
Start-up, shut-down and fault scenarios covered	
User friendly interfaces	
Lockable switch or password access	
Operator i/ps to be acknowledged	
Basic training specified	
Feature (SIL 2 and above)	**Evidence**
Protect against operator errors OR specify operator skill	
Feature (SIL 3)	**Evidence**
Protect against operator errors AND specify operator skill	
At least annual training	

Validation (Paras 7.3, 7.7, 7.9) (Tables A7, A9, B5, B8)

Feature (all SILs)	Evidence
Validation plan explaining technical and procedural steps including:Rev number, config management, when and who responsible, pass/fail, test environment, techniques (e.g. manual, auto, static, dynamic, statistical, computational)	
Plan reviewed	
Tests have chronological record	
Records and close-out report	
Calibration of equipment	
Suitable and justified choice of methods and models	
Feature (SIL 2 and above)	**Evidence**
Static analysis	
Test case metrics	
Feature (SIL 3)	**Evidence**
Simulation or modeling	
Further reviews (e.g. dead code, test coverage adequacy, behavior of algorithms) and traceability to the software design requirements	

Modifications (Para 7.8) Table A8

Feature (all SILs)	Evidence
Modification log	
Change control with adequate competence	
Software configuration management	
Impact analysis documented	
Re-verify changed modules	
Feature (SIL 2 and above)	**Evidence**
Re-verify affected modules	
Feature (SIL 3)	**Evidence**
Control of software complexity	
Re-validate whole system	

Acquired sub-systems

Feature (at the appropriate SIL)	Evidence
SIL requirements reflected onto suppliers	

Proven in use (Paras 7.4.2, 7.4.7)

Feature (at the appropriate SIL)	Evidence
Application appropriate	
Statistical data available	
Failure data validated	

Functional safety assessment (Para 8) (Tables A10, B4)

Feature (all SILs)	Evidence
Either checklists, truth tables, or block diagrams	
Feature (SIL 2 and above)	**Evidence**
As SIL 1	
Feature (SIL 3 and above)	**Evidence**
FMEA/Fault tree approach	
Common cause analysis of diverse software	

CHAPTER 5

Reliability Modeling Techniques

Chapter Outline

This chapter explains the techniques of quantified reliability prediction and are condensed from Reliability Maintainability and Risk, 8th Edition, David J Smith, Butterworth Heinemann (ISBN 978-0-08-096902-2).

5.1 Failure Rate and Unavailability

In Chapter 1, we saw that both failure rate (λ) and probability of failure on demand (PFD) are parameters of interest. Since unavailability is the probability of being failed at a randomly chosen moment then it is the same as the probability of failure on demand.

Safety Critical Systems Handbook. DOI: 10.1016/B978-0-08-096781-3.10005-7

PFD is dimensionless and is given by:

$$\text{PFD} = \text{UNAVAILABILITY} = (\lambda\,\text{MDT})/(1+\lambda\,\text{MDT}) \cong (\lambda\,\text{MDT})$$

where λ is failure rate and MDT is the mean down time (in consistent units). Usually λ MDT << 1.
For revealed failures the MDT consists of the active mean time to repair (MTTR) PLUS any logistic delays (e.g. travel, site access, spares procurement, administration). For unrevealed failures the MDT is related to the proof-test interval (T), PLUS the active MTTR, PLUS any logistic delays. The way in which failure is defined determines, to some extent, what is included in the down time. If the unavailability of a process is confined to failures whilst production is in progress then outage due to scheduled preventive maintenance is not included in the definition of failure. However, the definition of dormant failures of redundant units affects the overall unavailability (as calculated by the equations in the next Section).

5.2 Creating a Reliability Model

For any reliability assessment to be meaningful it is vital to address a specific system failure mode. Predicting the "spurious shutdown" frequency of a safety (shutdown) system will involve a different logic model and different failure rates from predicting the probability of "failure to respond".

To illustrate this, consider the case of a duplicated shutdown system whereby the voting arrangement is such that whichever sub-system recognizes a valid shutdown requirement then shutdown takes place (in other words "1 out of 2" voting).

When modeling the "failure to respond" event the "1 out of 2" arrangement represents redundancy and the two sub-systems are said to be "parallel" in that they both need to fail to cause the event. Furthermore the component failure rates used will be those which lead to ignoring a genuine signal. On the other hand, if we choose to model the "spurious shutdown" event the position is reversed and the sub-systems are seen to be "series" in that either failure is sufficient to cause the event. Furthermore the component failure rates will be for the modes which lead to a spurious signal.

The two most commonly used modeling methods are reliability block diagram analysis and fault tree analysis.

5.2.1 Block Diagram Analysis

Using the above example of a shut down system, the concept of a series reliability block diagram (RBD) applies to the "spurious shutdown" case.

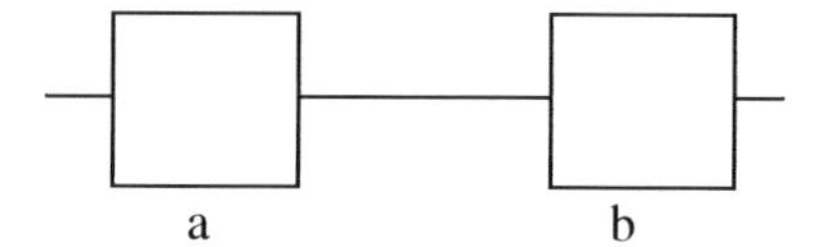

Figure 5.1: Series RBD.

The two sub-systems (a and b) are described as being "in series" since either failure causes the system failure in question. The mathematics of this arrangement is simple. We ADD the failure rates (or unavailabilities) of series items. Thus:

$$\lambda(\text{system}) = \lambda(\text{a}) + \lambda(\text{b})$$

and

$$\text{PFD(system)} = \text{PFD(a)} + \text{PFD(b)}$$

However, the "failure to respond" case is represented by the parallel block diagram model as follows:

The mathematics is dealt with in "Reliability Maintainability and Risk". However, the traditional results given prior to edition 7 of "Reliability Maintainability and Risk" and the majority of text books and standards have been challenged by K G L Simpson. It is now generally acknowledged that the traditional MARKOV model does not correctly represent the normal repair activities for redundant systems. The Journal of The Safety and Reliability Society, Volume 22, No 2, Summer 2002, published a paper by W G Gulland which agreed with those findings.

Tables 5.1 and 5.2 provide the failure rate and unavailability equations for simplex and parallel (redundant) identical sub-systems for revealed failures having a mean down time of MDT. However, it is worth mentioning that, as with all redundant systems, the total system failure rate (or PFD) will be dominated by the effect of common cause failure dealt with later in this chapter.

Unrevealed failures will eventually be revealed by some form of auto-test or proof-test. Whether manually scheduled or automatically initiated (e.g. auto-test using programmable

Table 5.1: System failure rates (revealed).

Number of Units	Number Required To Operate: 1	2	3	4
1	λ			
2	$2\lambda^2\text{MDT}$	2λ		
3	$3\lambda^3\text{MDT}^2$	$6\lambda^2\text{MDT}$	3λ	
4	$4\lambda^4\text{MDT}^3$	$12\lambda^3\text{MDT}^2$	$12\lambda^2\text{MDT}$	4λ

Table 5.2: System unavailabilities (revealed).

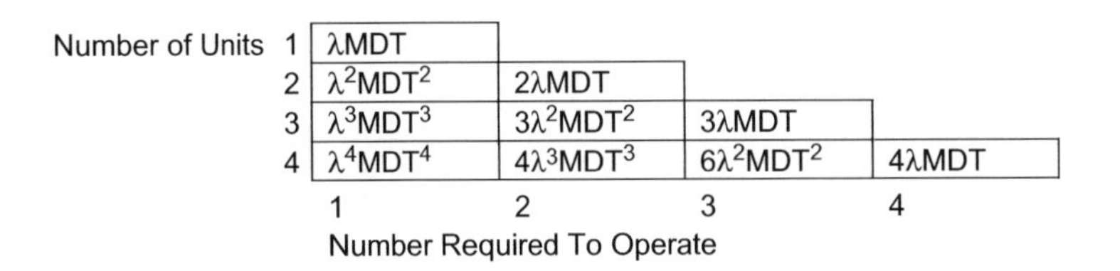

Number of Units	1	2	3	4
1	λMDT			
2	$\lambda^2 MDT^2$	$2\lambda MDT$		
3	$\lambda^3 MDT^3$	$3\lambda^2 MDT^2$	$3\lambda MDT$	
4	$\lambda^4 MDT^4$	$4\lambda^3 MDT^3$	$6\lambda^2 MDT^2$	$4\lambda MDT$

Number Required To Operate

logic) there will be a proof-test interval, T. Tables 5.3 and 5.4 provide the failure rate and unavailability equations for simplex and parallel (redundant) identical sub-systems for unrevealed failures having a proof test interval T. The MTTR is assumed to be negligible compared with T.

5.2.2 Common Cause Failure (CCF)

Whereas simple models of redundancy assume that failures are both random and independent, common cause failure (CCF) modeling takes account of failures which are linked, due to some dependency, and therefore occur simultaneously or, at least, within a sufficiently short interval as to be perceived as simultaneous.

Two examples are:

(a) the presence of water vapor in gas causing two valves to seize due to icing. In this case the interval between the two failures might be of the order of days. However, if the proof-test

Table 5.3: Failure rates (unrevealed).

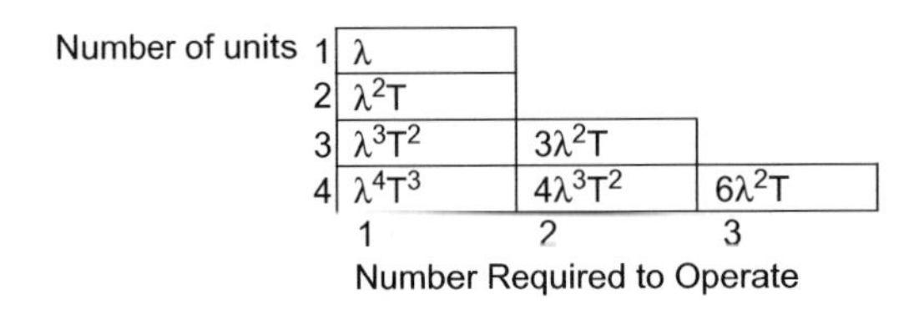

Number of units	1	2	3
1	λ		
2	$\lambda^2 T$		
3	$\lambda^3 T^2$	$3\lambda^2 T$	
4	$\lambda^4 T^3$	$4\lambda^3 T^2$	$6\lambda^2 T$

Number Required to Operate

Table 5.4: Unavailabilities (unrevealed).

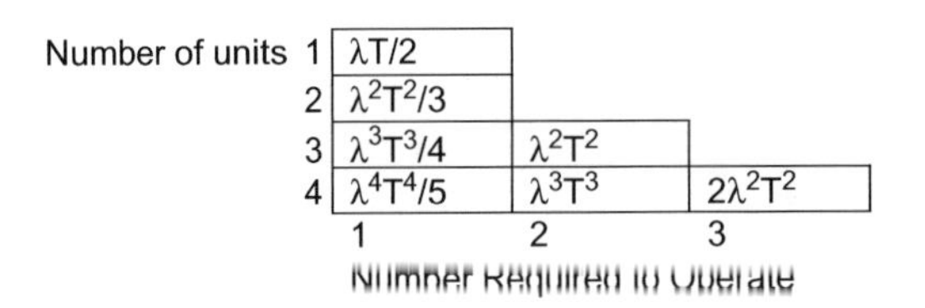

Number of units	1	2	3
1	$\lambda T/2$		
2	$\lambda^2 T^2/3$		
3	$\lambda^3 T^3/4$	$\lambda^2 T^2$	
4	$\lambda^4 T^4/5$	$\lambda^3 T^3$	$2\lambda^2 T^2$

Number Required to Operate

interval for this dormant failure is two months then the two failures will, to all intents and purposes, be simultaneous

(b) inadequately rated rectifying diodes on identical twin printed circuit boards failing simultaneously due to a voltage transient.

Typically, causes arise from

(a) Requirements: incomplete or conflicting
(b) Design: common power supplies, software, emc, noise
(c) Manufacturing: batch related component deficiencies
(d) Maintenance/operations: human induced or test equipment problems
(e) Environment: temperature cycling, electrical interference etc.

Defenses against CCF involve design and operating features which form the assessment criteria given in Appendix 3.

Common cause failures often dominate the unreliability of redundant systems by virtue of defeating the random coincident failure feature of redundant protection. Consider the duplicated system in Figure 5.2. The failure rate of the redundant element (in other words the coincident failures) can be calculated using the formula developed in Table 5.1, namely $2\lambda^2$MDT. Typical figures of 10 per million hours failure rate (10^{-5} per hr) and 24 hours down time lead to a failure rate of $2 \times 10^{-10} \times 24 = 0.0048$ per million hours. However, if only one failure in 20 is of such a nature as to affect both channels and thus defeat the redundancy, it is necessary to add the series element, shown as λ_2 in Figure 5.3, whose failure rate is $5\% \times 10^{-5} = 0.5$ per million hours, being two orders more frequent. The 5%, used in this example, is known as a BETA factor. The effect is to swamp the redundant part of the prediction and it is thus important to include CCF in reliability models. This sensitivity of system failure to CCF places emphasis on the credibility of CCF estimation and thus justifies efforts to improve the models.

In Figure 5.3, (λ_1) is the failure rate of a single redundant unit and (λ_2) is the common cause failure rate such that (λ_2) = β(λ_1) for the BETA model, which assumes that a fixed

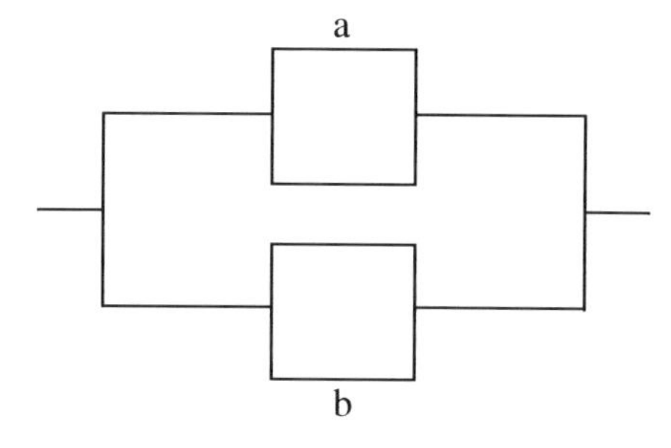

Figure 5.2: Parallel (redundant) RBD.

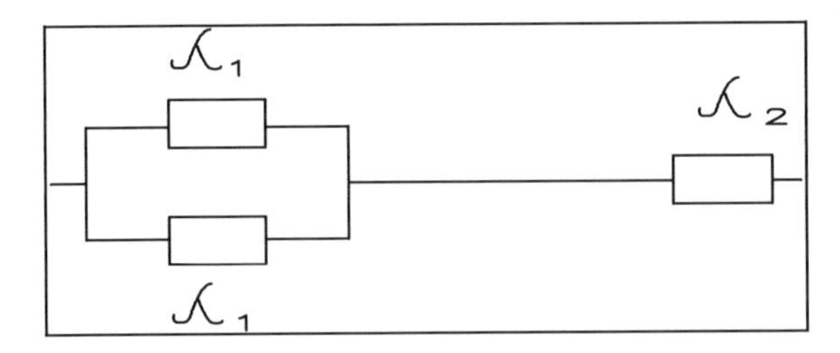

Figure 5.3: Reliability block diagram showing CCF.

proportion of the failures arise from a common cause. The contributions to BETA are split into groups of design and operating features which are believed to influence the degree of CCF. Thus the BETA multiplier is made up by adding together the contributions from each of a number of factors within each group. This Partial BETA model (as it is therefore known) involves the following groups of factors, which represent defenses against CCF:

Similarity (Diversity between redundant units reduces CCF)
Separation (Physical distance and barriers reduce CCF)
Complexity (Simpler equipment is less prone to CCF)
Analysis (FMEA and field data analysis will help to reduce CCF)
Procedures (Control of modifications and of maintenance activities can reduce CCF)
Training (Designers and maintainers can help to reduce CCF by understanding root causes)
Control (Environmental controls can reduce susceptibility to CCF, e.g. weather proofing of duplicated instruments)
Tests (Environmental tests can remove CCF prone features of the design, e.g. emc testing)

The Partial BETA model is assumed to be made up of a number of partial βs, each contributed to by the various groups of causes of CCF. β is then estimated by reviewing and scoring each of the contributing factors (e.g. diversity, separation).

The BETAPLUS model has been developed from the Partial Beta method because:

it is objective and maximizes traceability in the estimation of BETA. In other words the choice of checklist scores, when assessing the design, can be recorded and reviewed;
it is possible for any user of the model to develop the checklists further to take account of any relevant failure causal factors that may be perceived;
it is possible to calibrate the model against actual failure rates, albeit with very limited data;

there is a credible relationship between the checklists and the system features being analysed. The method is thus likely to be acceptable to the non-specialist;
the additive scoring method allows the partial contributors to β to be weighted separately;
the β method acknowledges a direct relationship between (λ_2) and (λ_1) as depicted in Figure 5.3;
it permits an assumed "non-linearity" between the value of β and the scoring over the range of β.

The BETAPLUS model includes the following enhancements:

(a) Categories of factors

Whereas existing methods rely on a single subjective judgement of score in each category, the BETAPLUS method provides specific design and operationally related questions to be answered in each category.

(b) Scoring

The maximum score for each question has been weighted by calibrating the results of assessments against known field operational data.

(c) Taking account of diagnostic coverage

Since CCF are not simultaneous, an increase in auto-test or proof-test frequency will reduce β since the failures may not occur at precisely the same moment.

(d) Sub-dividing the checklists according to the effect of diagnostics

Two columns are used for the checklist scores. Column (A) contains the scores for those features of CCF protection which are perceived as being enhanced by an increase in diagnostic frequency. Column (B), however, contains the scores for those features believed not to be enhanced by an improvement in diagnostic frequency. In some cases the score has been split between the two columns, where it is thought that some, but not all, aspects of the feature are affected (See Appendix 3).

(e) Establishing a model

The model allows the scoring to be modified by the frequency and coverage of diagnostic test. The (A) column scores are modified by multiplying by a factor (C) derived from diagnostic related considerations. This (C) score is based on the diagnostic frequency and coverage. (C) is in the range 1 to 3. A factor 'S', used to derive BETA, is then estimated from the RAW SCORE:

$$S = \text{RAW SCORE} = \left(\sum A \times C\right) + \sum B$$

(f) Non-linearity

There are currently no CCF data to justify departing from the assumption that, as BETA decreases (i.e. improves), then successive improvements become proportionately harder to achieve. Thus the relationship of the BETA factor to the RAW SCORE $[(\Sigma A \times C) + \Sigma B]$ is assumed to be exponential and this non-linearity is reflected in the equation which translates the raw score into a BETA factor.

(g) Equipment type

The scoring has been developed separately for programmable and non-programmable equipment, in order to reflect the slightly different criteria which apply to each type of equipment.

(h) Calibration

The model has been calibrated against field data.

Scoring criteria were developed to cover each of the categories (i.e. separation, diversity, complexity, assessment, procedures, competence, environmental control, environmental test). Questions have been assembled to reflect the likely features which defend against CCF. The scores were then adjusted to take account of the relative contributions to CCF in each area, as shown in the author's data. The score values have been weighted to calibrate the model against the data.

When addressing each question (in Appendix 3) a score less than the maximum of 100% may be entered. For example, in the first question, if the judgement is that only 50% of the cables are separated then 50% of the maximum scores (15 and 52) may be entered in each of the (A) and (B) columns (7.5 and 26).

The checklists are presented in two forms (listed in Appendix 3) because the questions applicable to programmable based equipments will be slightly different to those necessary for non-programmable items (e.g. field devices and instrumentation).

The headings (expanded with scores in Appendix 3) are:

(1) Separation/Segregation
(2) Diversity
(3) Complexity/Design/Application/Maturity/Experience
(4) Assessment/Analysis and Feedback of Data
(5) Procedures/Human Interface
(6) Competence/Training/Safety Culture
(7) Environmental Control
(8) Environmental Testing
Assessment of the diagnostic interval factor (C)

In order to establish the (C) score it is necessary to address the effect of diagnostic frequency. The diagnostic coverage, expressed as a percentage, is an estimate of the proportion of failures which would be detected by the proof-test or auto-test. This can be estimated by judgement or, more formally, by applying FMEA at the component level to decide whether each failure would be revealed by the diagnostics.

An exponential model is used to reflect the increasing difficulty in further reducing BETA as the score increases. This is reflected in the following equation which is developed in Smith D J, 2000, "Developments in the use of failure rate data":

$$\beta = 0.3 \exp(-3.4S/2624)$$

However, the basic BETA model applies to simple "one out of two" redundancy. In other words a pair of redundant items where the "top event" is the failure of both items. However, as the number of voted systems increases (in other words $N > 2$) the proportion of common cause failures varies and the value of β needs to be modified. The reason for this can be understood by thinking about two extreme cases:

1 out of 6

In this case only one out of the 6 items is required to work and up to 5 failures can be tolerated. Thus, in the event of a common cause failure, 5 more failures need to be provoked by the common cause. This is less likely than the "1 out of 2" case and β will be smaller. The table suggests a factor of 0.4.

5 out of 6

In this case 5 out of the 6 items are required to work and only 1 failure can be tolerated. Thus, in the event of a common cause failure, there are 5 items to which the common cause could apply. This is more likely than the "1 out of 2" case and β will be greater. The table suggests a factor of 8.

This is dealt with fully in the Manual of the Betaplus package. A portion of the table is shown as Table 5.5. IEC 61508 suggests slightly different values. This is an area of some debate, being based on intellectual reasoning rather than empirical data, and the "jury is still out".

Table 5.5: BETA(MooN) factor

	M=1	M=2	M=3	M=4
N=2	1			
N=3	0.3	2.4		
N=4	0.15	0.75	4	
N=5	0.075	0.45	1.2	6

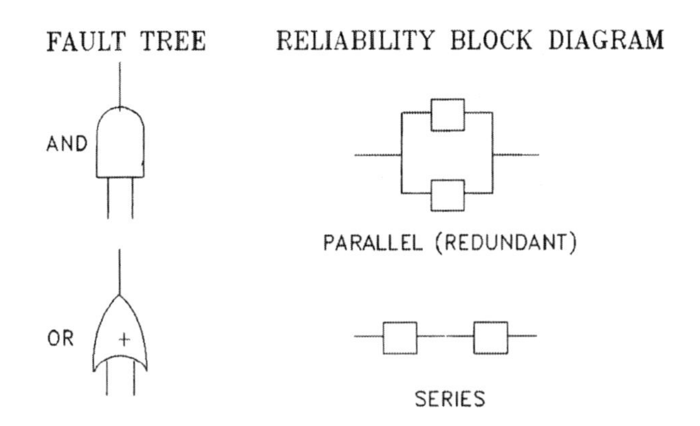

Figure 5.4: Series and Parallel equivalent to AND and OR.

5.2.3 Fault Tree Analaysis

Whereas the reliability block diagram provides a graphical means of expressing redundancy in terms of "parallel" blocks, fault tree analysis expresses the same concept in terms of paths of failure. The system failure mode in question is referred to as the Top Event and the paths of the tree represent combinations of event failures leading to the Top Event. The underlying mathematics is exactly the same. Figure 5.4 shows the OR gate which is equivalent to Figure 5.1 and the AND gate which is equivalent to Figure 5.2.

Figure 5.5 shows a typical fault tree modeling the loss of fire water arising from the failure of a pump, a motor, the detection or the combined failure of both power sources.

In order to allow for common cause failures in the fault tree model, additional gates are drawn as shown in the following examples. Figure 5.6 shows the reliability block diagram of Figure 5.3 in fault tree form.

The common cause failure can be seen to defeat the redundancy by introducing an OR gate along with the redundant G1 gate.

Figure 5.7 shows another example, this time of "2 out of 3" redundancy, where a voted gate is used.

5.3 Taking Account of Auto-test

The mean down time (MDT) of unrevealed failures is a fraction of the proof-test interval (i.e. for random failures, it is half the proof-test interval as far an individual unit is concerned) plus the actual MTTR (mean time to repair).

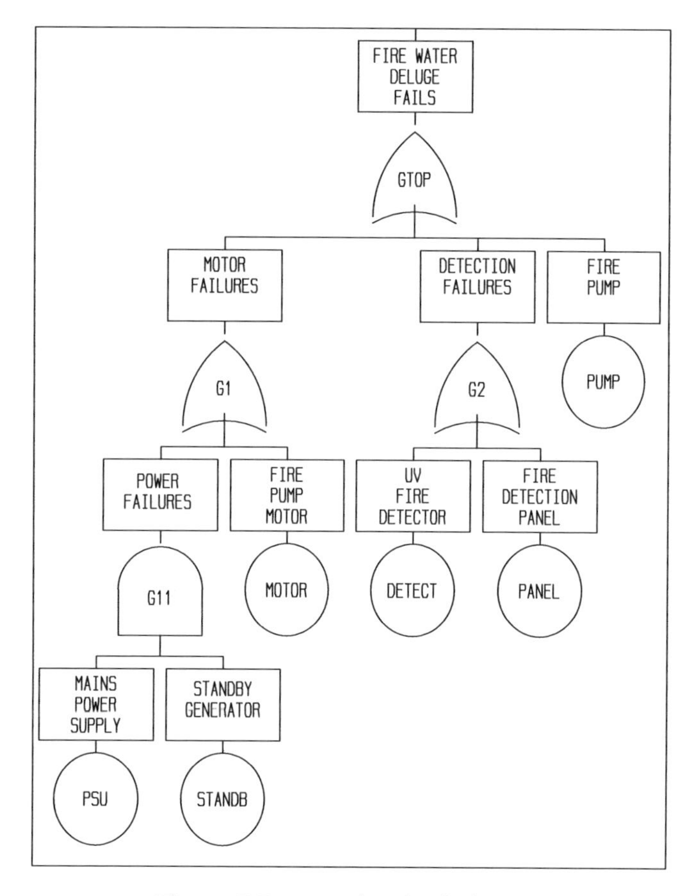

Figure 5.5: Example of a fault tree.

In many cases there is both auto-test, whereby a programmable element in the system carries out diagnostic checks to discover unrevealed failures, as well as a manual proof-test. In practice the auto-test will take place at some relatively short interval (e.g. 8 minutes) and the proof-test at a longer interval (e.g. one year).

The question arises as to how the reliability model takes account of the fact that failures revealed by the auto-test enjoy a shorter down time than those left for the proof-test. The ratio of one to the other is a measure of the diagnostic coverage and is expressed as a percentage of failures revealed by the test.

Consider now a dual redundant configuration (voted 1 out of 2) subject to 90% auto-test and the assumption that the manual test reveals 100% of the remaining failures.

The reliability block diagram needs to split the model into two parts in order to calculate separately in respect of the auto-diagnosed and manually-diagnosed failures.

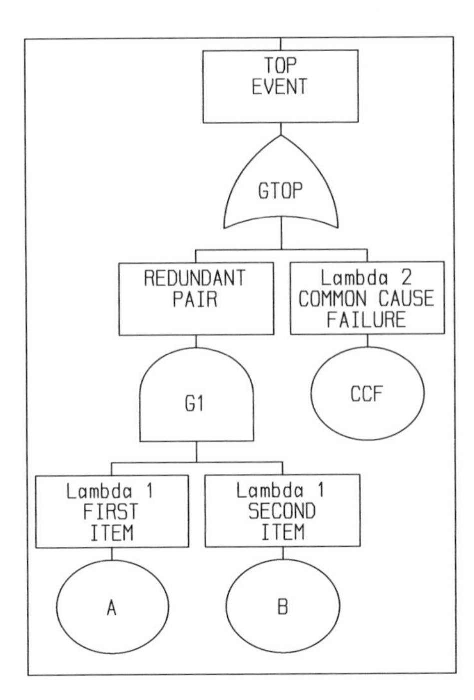

Figure 5.6: CCF in fault trees.

Figure 5.8 shows the parallel and common cause elements twice and applies the equations from Section 5.2 to each element. The failure rate of the item, for the failure mode in question, is λ. The equivalent fault tree is shown in Figure 5.9.

In IEC 61508 the following nomenclature is used to differentiate between failure rates which are either:

- Revealed or Unrevealed
- The failure mode in question or some other failure mode.

The term "dangerous failures" is coined for the "failure mode in question" and the practice has spread widely. It is, in the authors' opinion, slightly ambiguous. Whilst it is acknowledged that the term "dangerous" means in respect of the hazard being addressed, it nevertheless implies that the so-called "safe" failures are not hazardous. They may well be hazardous in some other respect.

The practice has become as follows:

λdd to mean failure rate of the revealed "dangerous failures"
λdu to mean failure rate of the unrevealed "dangerous failures"
λsd to mean failure rate of the revealed "safe failures"
λsu to mean failure rate of the unrevealed "safe failures"

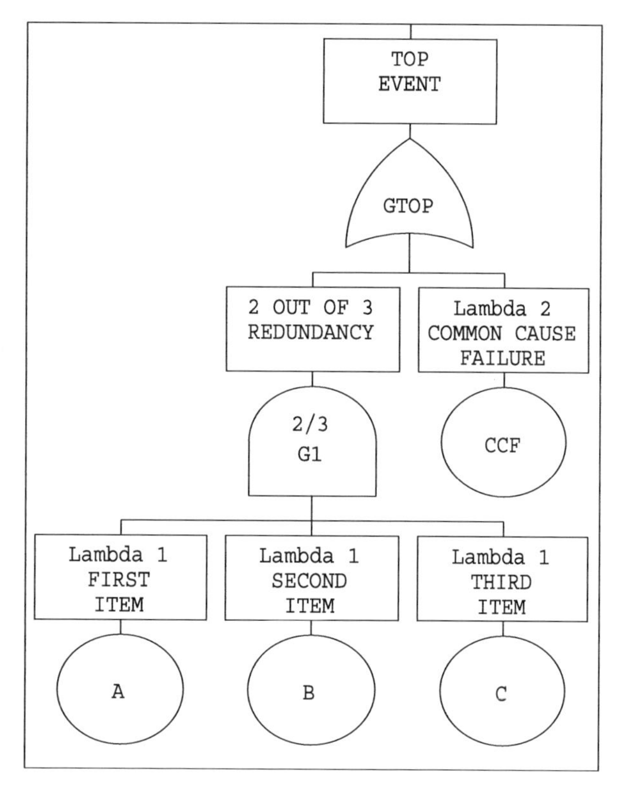

Figure 5.7: "2oo3" voting with CCF in a fault tree.

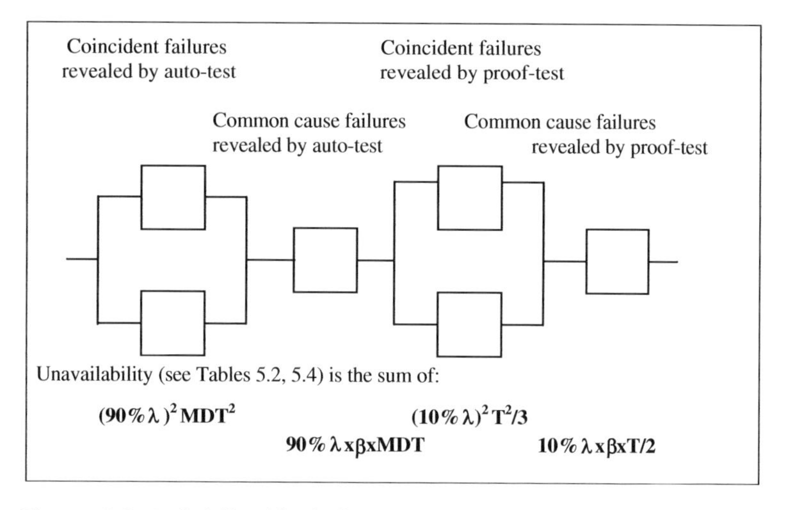

Figure 5.8: Reliability block diagram, taking account of diagnostics.

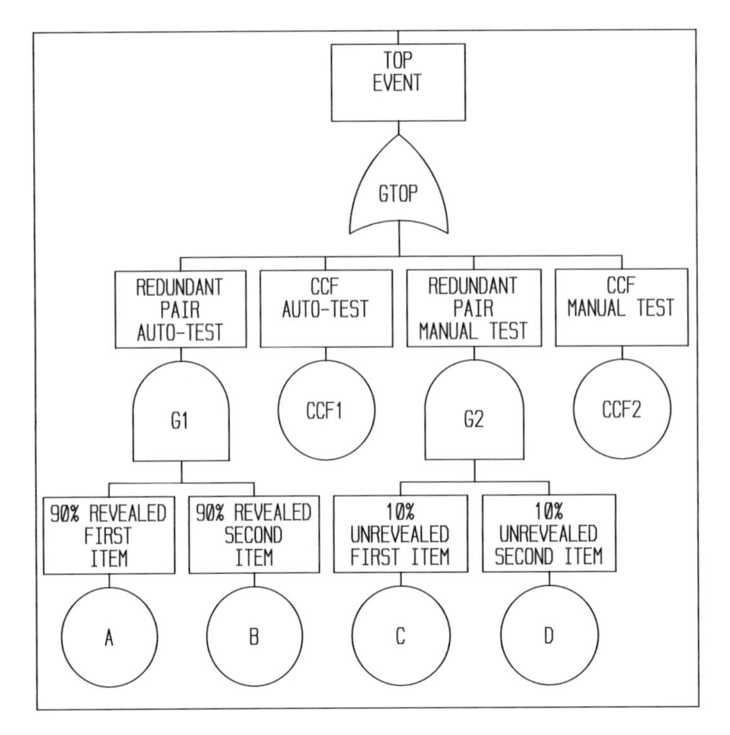

Figure 5.9: Fault tree diagram.

5.4 Human Factors

5.4.1 Addressing Human Factors

In addition to random coincident hardware failures, and their associated dependent failures (previous Section), it is frequently necessary to include human error in a prediction model (e.g. fault tree). Specific quantification of human error factors is not a requirement of IEC 61508. However, it is required that human factors are "considered".

It is well known that the majority of well-known major incidents, such as Three Mile Island, Bhopal, Chernobyl, Zeebrugge, Clapham and Paddington, are related to the interaction of complex systems with human beings. In short, the implication is that human error was involved, to a greater or lesser extent, in these and similar incidents. For some years there has been an interest in modeling these factors so that quantified reliability and risk assessments can take account of the contribution of human error to the system failure.

IEC 61508 (Part 1) requires the consideration of human factors at a number of places in the life-cycle. The assessment of human error is therefore implied. Table 5.6 summarizes the main references in the Standard.

Table 5.6: Human Factors References

Part 1		
Para 1.2	Scope	Makes some reference
Table 1	Life-cycle	Several uses of "to include human factors"
Para 7.3.2.1	Scope	Include humans
Para 7.3.2.5	Definition Stage	Human error to be considered
Para 7.4 various	Hazard/Risk Analysis	References to misuse and human intervention
Para 7.6.2.2	Safety Requirements Allocation	Availability of skills
Paras 7.7.2, 7.15.2	Ops & Maintenance	Refers to procedures
Part 2		
Para 7.4.10	Design and Development	Avoidance of human error
Para 7.6.2.3	Ops & Maintenance	Human error key element
Para 7.7.2.3	Validation	Includes procedures
Para 7.8.2.1	Modification	Evaluate mods on their effect on human interaction
Part 3		
Para 1.1	Scope	Human computer interfaces
Para 7.2.2.13	Specification	Human factors
Para 7.4.4.2	Design	Reference to Human error
Annex G	Data driven	Human factors

One example might be a process where there are three levels of defense against a specific hazard (e.g. over-pressure of a vessel). In this case the control valve will be regarded as the EUC. The three levels of defense are:

(1) The control system maintaining the setting of a control valve
(2) A shutdown system operating a separate shut-off valve in response to a high pressure
(3) Human response whereby the operator observes a high pressure reading and inhibits flow from the process.

The risk assessment would clearly need to consider how independent of each other are these three levels of protection. If the operator action (3) invokes the shutdown (2) then failure of that shutdown system will inhibit both defenses. In either case the probability of operator error (failure to observe or act) is part of the quantitative assessment.

Another example might be air traffic control, where the human element is part of the safety loop rather than an additional level of protection. In this case human factors are safety-critical rather than safety-related.

5.4.2 Human Error Rates

Human error rate data for various forms of activity, particularly in operations and maintenance, are needed. In the early 1960s there were attempts, by UKAEA, to develop a database of human error rates and these led to models of human error whereby rates could be estimated by assessing relevant factors such as stress, training and complexity. These human error probabilities include not only simple failure to carry out a given task, but diagnostic tasks where errors in reasoning, as well as action, are involved. There is not a great deal of data available due to the following problems:

- Low probabilities require large amounts of experience in order for meaningful statistics to emerge
- Data collection concentrates on recording the event rather than analysing the causes.
- Many large organizations have not been prepared to commit the necessary resources to collect data.

For some time there has been an interest in exploring the underlying reasons, as well as probabilities, of human error. As a result there are currently several models, each developed by separate groups of analysts working in this field. Estimation methods are described in the UKAEA document SRDA-R11, 1995. The better known are HEART (Human Error Assessment and Reduction Technique), THERP (Technique for Human Error Rate Prediction) and TESEO (Empirical Technique To Estimate Operator Errors).

For the earlier over-pressure example, failure of the operator to react to a high pressure (3) might be modeled by two of the estimation methods as follows:

"HEART" method

Basic task "Restore system following checks" — error rate = 0.003
Modifying factors:

Few independent checks	×3	50%
No means of reversing decision	×	25%

An algorithm is provided (not in the scope of this book) and thus:
Error probability = $0.003 \times [2 \times 0.5 + 1] \times [7 \times 0.25 + 1] = \mathbf{1.6 \times 10^{-2}}$

"TESEO" method

Basic task "Requires attention" – error rate = 0.01
$\times$ 1 for stress
$\times$ 1 for operator
$\times$ 2 for emergency
$\times$1 for ergonomic factors
Thus error probability = $0.01 \times 1 \times 1 \times 2 \times 1 = \mathbf{2 \times 10^{-2}}$

The two methods are in fair agreement and thus a figure of: 2×10^{-2} might be used for the example.

Figure 5.10 shows a fault tree for the example assuming that the human response is independent of the shutdown system. The fault tree models the failure of the two levels of protection (2) and (3). Typical (credible) probabilities of failure on demand are used for the initiating events. The human error value of 2×10^{-2} could well have been estimated as above.

Quantifying this tree would show that the overall probability of failure on demand is 1.4×10^{-4} (incidentally meeting SIL 3 quantitatively).

Looking at the relative contribution of the combinations of initiating events would show that human error is involved in over 80% of the total. Thus, further consideration of human error factors would be called for.

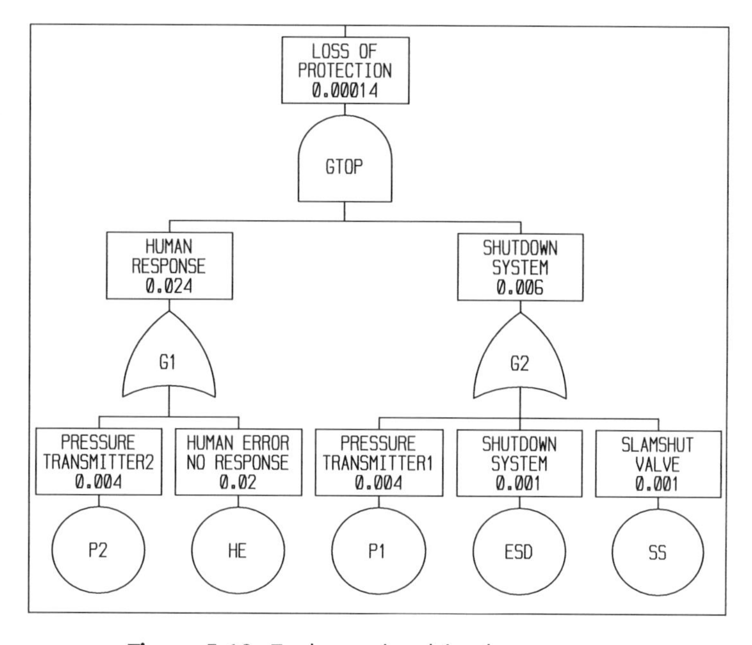

Figure 5.10: Fault tree involving human error.

5.4.3 A Rigorous Approach

There is a strong move to limit the assessment of human error probabilities to 10^{-1} unless it can be shown that the human action in question has been subject to some rigorous review. The HSE have described a seven step approach which involves:

STEP 1 Consider main site hazards
e.g. A site HAZOP identifies the major hazards.
STEP 2 Identify manual activities that effect these hazards
The fault tree modeling of hazards will include the human errors which can lead to the top events in question.
STEP 3 Outline the key steps in these activities
Task descriptions, frequencies, task documentation, environmental factors and competency requirements.
STEP 4 Identify potential human failures in these steps
The HEART and TESEO methodologies can be used as templates to address the factors.
STEP 5 Identify factors that make these failures more likely
Review the factors which contribute (The HEART list is helpful)
STEP 6 Manage the failures using hierarchy of control
Can the hazard be removed, mitigated etc.
STEP 7 Manage Error Recovery
Involves alarms, responses to incidents etc.

Anecdotal data as to the number of actions, together with the number of known errors, can provide estimates for comparison with the HEART and TESEO predictions. Good agreement between the three figures helps to build confidence in the assessment.

CHAPTER 6

Failure Rate and Mode Data

Chapter Outline

In order to quantify reliability models it is necessary to obtain failure rate and failure mode data.

6.1 Data Accuracy

There are many collections of failure rate data compiled by defense, telecommunications, process industries, oil and gas and other organizations. Some are published Data Handbooks such as:

US MIL HANDBOOK 217 (Electronics)
CNET (French PTT) Data
HRD (Electronics, British Telecom)
RADC Non-Electronic Parts Handbook NPRD
OREDA (Offshore data)
FARADIP.THREE (Data ranges)

Some are data banks which are accessible by virtue of membership or consultancy fee such as:

SRD (Part of UKAEA) Data Bank
Technis (Tonbridge)

Safety Critical Systems Handbook. DOI: 10.1016/B978-0-08-096781-3.10006-9

Some are in-house data collections which are not generally available. These occur in:

Large industrial manufacturers
Public utilities.

These data collection activities were at their peak in the 1980s but, sadly, many declined during the 1990s and many of the published sources have not been updated since that time.

Failure data are usually, unless otherwise specified, taken to refer to random failures (i.e. constant failure rates). It is important to read, carefully, any covering notes since, for a given temperature and environment, a stated component, despite the same description, may exhibit a wide range of failure rates because:

- Some failure rate data include items replaced during preventive maintenance whereas others do not. These items should, ideally, be excluded from the data but, in practice, it is not always possible to identify them. This can affect rates by an order of magnitude.
- Failure rates are affected by the tolerance of a design. Because definitions of failure vary, a given parametric drift may be included in one data base as a failure, but ignored in another. This will cause a variation in the values.
- Although nominal environmental and quality assurance levels are described in some databases, the range of parameters covered by these broad descriptions is large. They represent, therefore, another source of variability.
- Component parts are often only described by reference to their broad type (e.g. signal transformer). Data are therefore combined for a range of similar devices rather than being separately grouped, thus widening the range of values. Furthermore, different failure modes are often mixed together in the data.
- The degree of data screening will affect the relative numbers of intrinsic and induced failures in the quoted failure rate.
- Reliability growth occurs where field experience is used to enhance reliability as a result of modifications. This will influence the failure rate data.
- Trial and error replacement is sometimes used as a means of diagnosis and this can artificially inflate failure rate data.
- Some data record undiagnosed incidents and "no fault found" visits. If these are included in the statistics as faults, then failure rates can be inflated.

Quoted failure rates are therefore influenced by the way they are interpreted by an analyst and can span one or two orders of magnitude as a result of different combinations of the above factors. **Prediction calculations were explained in Chapter 5 and it will be seen that the relevance of failure rate data is more important than refinements in the statistics of the calculation.** Data sources can at least be subdivided into "site specific", "industry specific" and "generic" and work has shown (Smith D J, 2000, Developments in the use of failure rate data [illegible]) that the more specific the data source the greater the confidence in the prediction.

Failure rates are often tabulated, for a given component type, against ambient temperature and the ratio of applied to rated stress (power or voltage). Data are presented in one of two forms:

- *Tables:* lists of failure rates, with or without multiplying factors, for such parameters as quality and environment.
- *Models:* obtained by regression analysis of the data. These are presented in the form of equations which yield a failure rate as a result of inserting the device parameters into the appropriate expression.

Because of the large number of variables involved in describing microelectronic devices, data are often expressed in the form of models. These regression equations (WHICH GIVE A TOTALLY MISLEADING IMPRESSION OF PRECISION) involve some or all of the following:

Complexity (number of gates, bits, equivalent number of transistors)
Number of pins
Junction temperature
Package (ceramic and plastic packages)
Technology (CMOS, NMOS, bipolar, etc.)
Type (memory, random LSI, analogue, etc.)
Voltage or power loading
Quality level (affected by screening and burn-in)
Environment
Length of time in manufacture.

Although empirical relationships have been established relating certain device failure rates to specific stresses, such as voltage and temperature, no precise formula exists which links specific environments to failure rates. The permutation of different values of environmental factors is immense. General adjustment (multiplying) factors have been evolved and these are often used to scale up basic failure rates to particular environmental conditions.

Because Failure Rate is, probably, the least precise engineering parameter, it is important to bear in mind the limitations of a Reliability prediction. The work mentioned above (Smith DJ, 2000) makes it possible to **express predictions using confidence intervals**. The resulting MTBF, Availability (or other parameter), should not be taken as an absolute value but rather as a general guide to the design reliability. Within the prediction, however, the relative percentages of contribution to the total failure rate are of a better accuracy and provide a valuable tool in design analysis.

Owing to the differences between data sources, comparisons of reliability should always involve the same data source in each prediction.

For any reliability assessment to be meaningful it must address a specific system failure mode. To predict that a safety (shutdown) system will fail at a rate of, say, once per annum is, on its own, saying very little. It might be that 90% of the failures lead to a spurious shutdown and 10% to a failure to respond. If, on the other hand, the ratios were to be reversed then the picture would be quite different.

The failure rates, mean times between failures or availabilities must therefore be assessed for defined failure types (modes). In order to achieve this, the appropriate component level failure modes must be applied to the prediction models which were described in Chapter 5. Component failure mode data are sparse but a few of the sources do contain some information. The following Paragraphs indicate where this is the case.

6.2 Sources of Data

Sources of failure rate and failure mode data can be classified as:

- Site specific

  Failure rate data which have been collected from similar equipment being used on very similar sites (e.g. two or more gas compression sites where environment, operating methods, maintenance strategy and equipment are largely the same). Another example would be the use of failure rate data from a flow corrector used throughout a specific distribution network. This data might be applied to the RAMS (reliability, availability, maintainability, safety) prediction for a new design of circuitry for the same application.
- Industry specific

  An example would be the use of the OREDA offshore failure rate data book for a RAMS prediction of a proposed offshore process package.
- Generic

  A generic data source combines a large number of applications and sources (e.g. FARADIP.THREE).

As has already been emphasized, predictions require failure rates for specific modes of failure (e.g. open circuit, signal high, valve closes). Some, but unfortunately only a few, data sources contain specific failure mode percentages. Mean time to repair data is even more sparse although the OREDA data base is very informative in this respect.

The following are the more widely used sources.

6.2.1 Electronic Failure Rates

- US Military Handbook 217 (Generic, no failure modes)
- HRD5 Handbook of Reliability Data for Electronic Components Used in Telecommunications Systems (Industry specific, no failure modes)

- Recueil de Donnés de Fiabilité du CNET (Industry specific, no failure modes)
- BELLCORE, (Reliability Prediction Procedure for Electronic Equipment) TR-NWT-000332 Issue 5 1995 (Industry specific, no failure modes)
- Electronic data NOT available for purchase

A number of companies maintain failure rate data banks, including Nippon Telephone Corporation (Japan), Ericson (Sweden) and Thomson CSF (France) but these data are not generally available outside the organizations.

6.2.2 Other General Data Collections

- Nonelectronic Parts Reliability Data Book – NPRD (Generic, Some failure modes)
- OREDA - Offshore Reliability Data (1984/92/95/97/02) (Industry specific, Detailed failure modes, Mean times to repair)
- FARADIP.THREE (the author) (Industry and generic, many failure modes, some repair times)
- UKAEA (Industry and generic, many failure modes)
- Sources of Nuclear Generation Data (Industry specific)

In the UKAEA documents, above, there are some nuclear data, as in NNC (National Nuclear Corporation), although this may not be openly available.

In the USA, Appendix III of the WASH 1400 study provided much of the data frequently referred to and includes failure rate ranges, event probabilities, human error rates and some common cause information. The IEEE Standard IEEE500 also contains failure rates and restoration times. In addition there is NUCLARR (Nuclear Computerized Library for Assessing Reliability) which is a PC based package developed for the Nuclear Regulatory Commission and containing component failure rates and some human error data. Another US source is the NUREG publication. Some of the EPRI data are related to nuclear plant. In France, Electricity de France provides the EIReDA mechanical and electrical failure rate data base which is available for sale. In Sweden the TBook provides data on components in Nordic Nuclear Power Plants.

- US Sources of Power Generation Data (Industry specific)

The EPRI (Electrical Power Research Institute) of GE Co., New York, data scheme is largely gas turbine generation failure data in the USA.

There is also the GADS (Generating Availability Data System) operated by NERC (North American Electric Reliability Council). They produce annual statistical summaries based on experience from power stations in USA and Canada.

- SINTEF (Industry specific).

SINTEF is the Foundation for Scientific and Industrial Research at the Norwegian Institute of Technology. They produce a number of reliability handbooks which include failure rate data for various items of process equipment.

- Data not available for purchase

Many companies (e.g. Siemens), and for that matter firms of RAMS consultants (e.g. RM Consultants Ltd), maintain failure rate data but only for use by that organization.

6.2.3 Some Older Sources

A number of sources have been much used and are still frequently referred to. They are, however, somewhat dated but are listed here for completeness.

Reliability Prediction Manual for Guided Weapon Systems (UK MOD) – DX99/013-100
Reliability Prediction Manual for Military Avionics (UK MOD) – RSRE250
UK Military Standard 00-41
Electronic Reliability Data – INSPEC/NCSR (1981)
Green and Bourne (book), Reliability Technology, Wiley 1972
Frank Lees (book), Loss Prevention in the Process Industries, Butterworth Heinemann.

6.2.4 Manufacturer's Data

There is a rapidly increasing trend to quote failure rates offered by equipment manufacturers. Extreme care should be exercised in the use of such failure rate data. Only users can claim to record all failures. There are numerous reasons why these failure rates can be highly optimistic. Reasons include:

- Items in store before use
- Items still in the supply chain
- Failed item tolerated due to replacement causing process disruption and the ability to continue in degraded mode due to information redundancy
- Item replaced by user w/o returning
    - Disillusioned by supplier
    - Not worth the cost (low value item)
    - No warranty incentive
    - Feedback not encouraged
    - User fixes it
- Transient fault subsequently appears as no "fault found"
- Mismatch between perceived calendar vs operating hours for the item (standby items etc.)
- Failure discounted due to inappropriate environment despite the fact that real-life failure rates include these
- Vested interest in optimism
- The data were actually only a reliability prediction.

Technis studies indicate that manufacturer's data can be up to an order of magnitude optimistic (on average 5:1).

6.2.5 Anecdotal Data

Although not as formal as data based on written maintenance records, this important source should not be overlooked. Quantities of failures quoted by long-serving site personnel are likely to be fairly accurate and might even, in some cases, be more valuable than records-based data. The latter pass from maintainer to record keeper to analyst and may lose accuracy due to interpretation through the chain of analysis. Anecdotal data, on the other hand, can be challenged and interpreted first hand.

6.3 Data Ranges and Confidence Levels

For some components there is fairly close agreement between the sources and in other cases there is a wide range, the reasons for which were summarized above. For this reason predictions are subject to wide tolerances.

The ratio of predicted failure rate (or system unavailability) to field failure rate (or system unavailability) was calculated for each of 44 examples and the results (see Smith DJ, 2000) were classified under the three categories described in Section 6.2, namely:

Predictions using site specific data
Predictions using industry specific data
Predictions using generic data.

The results are:

For a Prediction Using Site Specific Data

One can be this confident	That the eventual field failure rate will be BETTER than:
95%	3½ times the predicted
90%	2½ times the predicted
60%	1½ times the predicted

For a Prediction Using Industry Specific Data

One can be this confident	That the eventual field failure rate will be BETTER than:
95%	5 times the predicted
90%	4 times the predicted
60%	2½ times the predicted

For a Prediction Using Generic Data

One can be this confident	That the eventual field failure rate will be BETTER than:
95%	8 times the predicted
90%	6 times the predicted
60%	3 times the predicted

Additional evidence in support of the 8:1 range is provided from the FARADIP.THREE data bank, which shows an average of 7:1 across the ranges.

The FARADIP.THREE data base was created to show the ranges of failure rate for most component types. This database, currently version 6.5 in 2010, is a summary of most of the other databases and shows, for each component, the range of failure rate values which is to be found from them. Where a value in the range tends to predominate then this is indicated. Failure mode percentages are also included. It is available on disk from Technis at 26 Orchard Drive, Tonbridge, Kent TN10 4LG, UK (technis.djs@virgin.net) and includes:

- Discrete
  - Diodes
  - Opto-electronics
  - Lamps and displays
  - Crystals
  - Tubes
- Passive
  - Capacitors
  - Resistors
  - Inductive
  - Microwave
- Instruments and Analysers
  - Analysers
  - Fire and Gas detection
  - Meters
  - Flow instruments
  - Pressure instruments
  - Level instruments
  - Temperature instruments
- Connection
  - Connections and connectors
  - Switches and breakers
  - PCBs cables and leads
- Electro-mechanical
  - Relays and solenoids
  - Rotating machinery (fans, motors, engines)
- Power
  - Cells and chargers
  - Supplies and transformers
- Mechanical
  - Pumps
  - Valves and valve parts
  - Leakage
  - Bearings
  - Miscellaneous
- Pneumatics
- Hydraulics
- Computers, data processing and communications
- Alarms, fire protection, arresters and fuses

6.4 Conclusions

The use of stress-related regression models implies an unjustified precision in estimating the failure rate parameter.

Site specific data should be used in preference to industry specific data which, in turn, should be used in preference to generic data.

Predictions should be expressed in confidence limit terms using the above information. The warnings concerning the optimism of manufacturer's data should be borne in mid.

In practice, failure rate is a system level effect. It is closely related to, but not entirely explained by, component failure. A significant proportion of failures encountered with modern electronic systems are not the direct result of parts failures but of more complex interactions within the system. The reason for this lack of precise mapping arises from such effects as human factors, software, environmental interference, interrelated component drift and circuit design tolerance.

The primary benefit to be derived from reliability and safety engineering is the reliability and integrity growth which arises from ongoing analysis and follow-up as well as from the corrective actions brought about by failure analysis. Reliability prediction, based on the manipulation of failure rate data, involves so many potential parameters that a valid repeatable model for failure rate estimation is not possible. Thus, failure rate is the least accurate of engineering parameters and prediction from past data should be carried out either:

As an indicator of the approximate level of reliability of which the design is capable, given reliability growth in the field
To provide relative comparisons in order to make engineering decisions concerning optimum redundancy
As a contractual requirement
In response to safety-integrity requirements.

It should not be regarded as an exact indicator of future field reliability as a result of which highly precise prediction methods are often, by reason of poor data, not justified.

Now try the exercise and the example, which are Chapters 11 and 12.

CHAPTER 7

Demonstrating and Certifying Conformance

Chapter Outline

7.1 Demonstrating Conformance

It is becoming increasingly necessary to demonstrate (or even certify) conformance to the requirements of IEC61508. This has been driven by customer demands for certification coupled with suppliers' aspirations not to be "left out" of the trend. There are two types of certification.

> FIRSTLY: That an organization can demonstrate the generic capability to produce such a product or system (i.e. that it has the necessary procedures and competence in place).
> SECONDLY: That a specific product or system design meets the requirements outlined in the preceding chapters (i.e. that the above procedures have been implemented).

In the first case it is the raft of procedures and work practices, together with the competence of individuals, which is being assessed. This is known as the Functional Safety Capability (FSC) of an organization and is now more commonly referred to as Functional Safety Management (FSM). It is demonstrated by an appropriate quality management system and evidenced by documented audits and examples of the procedures being used.

In the second it is the design and the life-cycle activities of a particular product which are being assessed. This is demonstrated by specifications, design documents, reviews, test specifications and results, failure rate predictions, FMEAs to determine safe failure fraction and so on.

Safety Critical Systems Handbook. DOI: 10.1016/B978-0-08-096781-3.10007-0

In practice, however, it is not really credible to assess one of the above without evidence of the other. FSM needs to be evidenced by at least one example of a product or project and a product's conformance needs to be evidenced by documentation and life-cycle activities which show overall capability.

7.2 The Current Framework for Certification

Most people in industry are, by now, well aware of the certification framework for ISO 9001. UKAS (The United Kingdom Accreditation Service) ***accredits*** organizations to be able to ***certify*** clients to ISO 9001.

There are over 100 accredited bodies (in the UK alone) offering ISO 9001 certification and many thousands of organizations who have been certified, by them, to the ISO 9001 standard. There are only two outcomes – one either meets the standard or one does not.

The situation for IEC 61508 is rather different and less well developed.

Firstly, as explained above, there are the two aspects to the certification (namely the organization or the product). Unlike ISO 9001, there are four levels of rigor against which to be certified (SILs 1–4). In addition, for the organization, the certificate will be granted for a specific scope such as supply to certain industry sectors, technologies used, life-cycle phases, etc.

Following a DTI initiative in 1998/9, a framework was developed by CASS Ltd (Conformity Assessment of Safety-related Systems). One motive for this was to erode differences in approach across application sectors and thereby improve the marketability of UK safety-related technology. Another was to prevent multiple assessments and also to meet the need for the ever increasing demand for assessment of safety-related equipment. The CASS framework suggested five types of assessment. In the fullness of time this has developed as two types.

- Functional Safety Capability (or Management) Assessment (known as FSCA, or FSM). Described in Chapter 2 and catered for by Appendix 1 of this book
- Specific Product/Systems Assessment. This is the overall assessment of whether a system meets specific SIL targets, as addressed throughout this book

At present UKAS (United Kingdom Accreditation Service) have accredited two bodies (Figure 7.1):

SIRA Certification Service to certify
- Functional Safety Capability
- Products/Systems Hardware
- Products/Systems Software

BASEEFA Ltd to certify
- Products/Systems Hardware

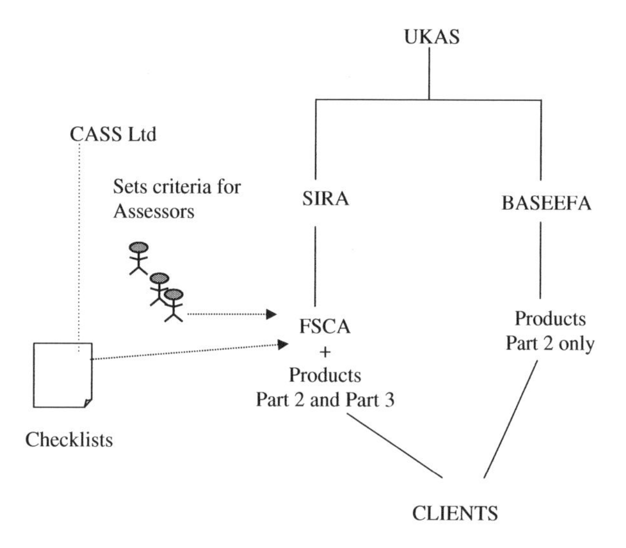

Figure 7.1: Certification framework.

There are other certification bodies emerging (not necessarily accredited by UKAS). It is not possible to give a detailed list in a book of this type, due to the rapidly changing situation.

7.3 Self Certification (Including Some Independent Assessment)

There is nothing to prevent self-assessment, either of one's Functional Safety Capability, as an organization, or of the Safety Integrity Level of a product or design. Indeed this can be, and often is, as rigorous as the accredited certification process.

Third-party involvement in the assessment, whilst not essential, is desirable to demonstrate impartiality and one requires a safety professional specializing in this field. The Safety and Reliability Society, which is associated with the Engineering Council, maintains appropriate standards for admission to corporate membership and membership would be one factor in suggesting suitability. Suitable consultants should have dealt with many other clients and have a track record concerning IEC 61508. Examples would be papers, lectures, assessments and contributions to the drafting of the standard. This would serve to demonstrate that some assessment benchmark has been applied.

As a minimum self-assessment requires:

7.3.1 Showing Functional Safety Capability (FSM) as Part of the Quality Management System

This is described in Chapter 2, being one of the requirements of Part 1 of IEC 61508. Appendix 1 of this book provides a template procedure which could be tailored and integrated into an organization's quality management system.

The organization would show evidence of both audits and reviews of the procedure in order to claim compliance. Compliance with ISO 9001 is strongly indicated if one is aiming to claim functional safety compliance. The life-cycle activities are so close to the ISO 9001 requirements that it is hard to imagine a claim which does not include them. The ISO 9001 quality management system would need to be enhanced to include:

Safety-related competencies (see Chapter 2)
Functional safety activities (Appendix 1)
Techniques for (and examples of) assessment (Chapters 5 and 6).

The scope of the capability should also be carefully defined because no one organization is likely to be claiming to perform every activity described in the life-cycle. Examples of scope might include:

Design and build of safety-related systems
Design and build of safety-related instrumentation
Assessment of SIL targets and of compliance of systems
Maintenance of safety-related equipment.

7.3.2 Application of IEC 61508 to Projects/Products

In addition to the procedural capability described in Section 7.3.1 a self-assessment will also need to demonstrate the completion of at least one project together with a safety-integrity study.

The tables at the end of Chapters 3, 4 and 8 of this book provide a means of formally recording the reviews and assessments. Chapters 11, 12, 14 and 16 show examples of how the quantitative assessments can be demonstrated.

7.3.3 Rigor of Assessment

In addition to the technical detail suggested by Section 7.3.2 above, there needs to be visible evidence that sufficient aspects of assessment have been addressed. The "assessment schedule" checklist in Appendix 2 of this book provides a formal checklist which allows one to demonstrate the thoroughness (i.e. rigor) of an assessment.

7.3.4 Independence

This has been covered in Chapter 1.4 and the same provisions apply.

It has to be acknowledged that third-party assessment does involve additional cost for perhaps little significant added value in terms of actual safety-integrity. Provided that the self-assessments are conducted under a formal quality management system, with appropriate audits, and provided also that competency of the assessors in risk assessment can be

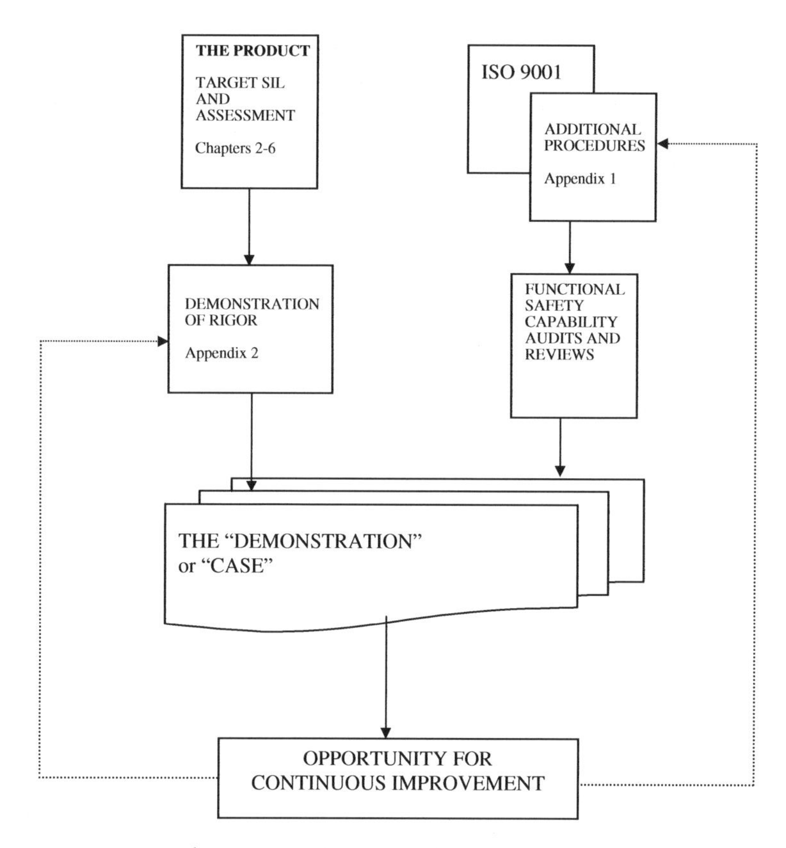

Figure 7.2: Elements of self assessment.

demonstrated by the organization, then there is no reason why such assessment should not be both credible and thus acceptable to clients and regulators.

Clearly, some evidence of external involvement in the setting up and periodic auditing of self assessment schemes will enhance this credibility provided that the external consultant or organization can demonstrate sufficient competence in this area.

Proactive involvement in professional institutions, industrial research organizations or the production and development of IEC 61508 and associated standards by both self-assessors and external consultants would assist in this respect. The authors, for example, have made major contributions to the Standard and to a number of the second-tier documents described in Chapters 8–10. Thus, the credibility of third-party assessment bodies or consultants does need to be addressed vigorously.

Figure 7.2 shows how a "Demonstration of Conformance" might be built up from the elements described in this chapter. This "Demonstration" would provide backup to any safety report where a level of safety-integrity is being claimed. It also provides a mechanism for continuous improvement as suggested by the assessment techniques themselves.

7.4 Preparing for Assessment

Whether the assessment is by an accredited body (e.g. SIRA) or a third-party consultant, it is important to prepare in advance. The assessor does not know what you know and, therefore, the only visibility of your conformance is provided by documented evidence of:

Functional safety procedures
Specifications
Audits against procedures
Reviews of the adequacy of procedures
Design reviews of projects
Test plans, reports and remedial action
Safety-integrity assessments
Competency register.

A visible trail of reviews, whereby the procedures and work practices have been developed in practice, is a good indicator that your organization is committed to Functional Safety.

Being ill-prepared for an assessment is very cost-ineffective. Man-hours and fees are wasted on being told what a simple internal audit could have revealed.

The majority of assessments are based on the method of:

A pre-assessment to ascertain whether the required procedures and practices are in place (often referred to as gap-analysis)
A final assessment where the procedures are reviewed in detail and evidence is sought as to their implementation.

With sensible planning these stages can be prepared for in advance and the necessary reviews conducted internally. It is important that evidence is available to assessors for all the elements of the life-cycle

Assessments may result in:

Major non-compliances
Minor non-compliances
Observations.

A major non-compliance would arise if a life-cycle activity is clearly not evidenced. For example, the absence of any requirement for assessment of safe failure fraction would constitute a major non-compliance with the Standard. More than one major non-compliance would be likely to result in the assessment being suspended until the client declared himself ready for re-assessment. This would be unnecessarily expensive when the situation could be prevented by adequate preparation.

A minor non-compliance might arise if an essential life-cycle activity, although catered for in the organization's procedures, has been omitted. For example a single project where there were inadequate test records would attract a minor non-compliance.

Observations might include comments of how procedures might be enhanced.

7.5 Summary

It is important to ensure that any product assessment concentrates primarily on the technical aspects of a safety-related system. In other words it should address all the aspects (quantitative and qualitative) described in this book. Product assessment (and potentially certification) is currently offered at two levels:

- The random hardware failures and Safe failure fraction only
- All aspects (the 7 steps in Chapter 1) including life-cycle activities.

The latter is, of course, a more substantial form of demonstration but requires considerably more resources and hence cost. The trend, in the case of accredited certification, is towards the fuller demonstration.

Procedures and document hierarchies are important, of course, for without them the technical assessment would have no framework upon which to exist and no visibility to demonstrate its findings. However, there is a danger that a "blinkered attention to detail" approach can concentrate solely on the existence of procedures and of specific document titles. Procedures, and the mere existence of documents, do not of themselves imply achieved functional safety unless they result in technical activity.

Similarly, documents alone do not enhance function safety; they are a vehicle to implement technical requirements. Their titles are relatively unimportant and it is necessary to see behind them to assess whether the actual requirements described in this book have been addressed and implemented. The same applies to safety management systems generally.

If this is borne in mind then assessment, be it self generated or third party, can be highly effective.

PART B

Specific Industry Sectors

Some of the following documents are referred to as "second tier" guidance in relation to IEC 61508. Due to the open ended nature of the statements made, and to ambiguity of interpretation, it cannot be said that conformance with any one of them automatically implies compliance with IEC 61508.

However, they cover much the same ground as each other albeit using slightly different terms to describe documents and life-cycle activities.

Figure B.1 illustrates the relationship of the documents to IEC 61508. A dotted line indicates that the document addresses similar issues whilst not strictly being viewed as second tier.

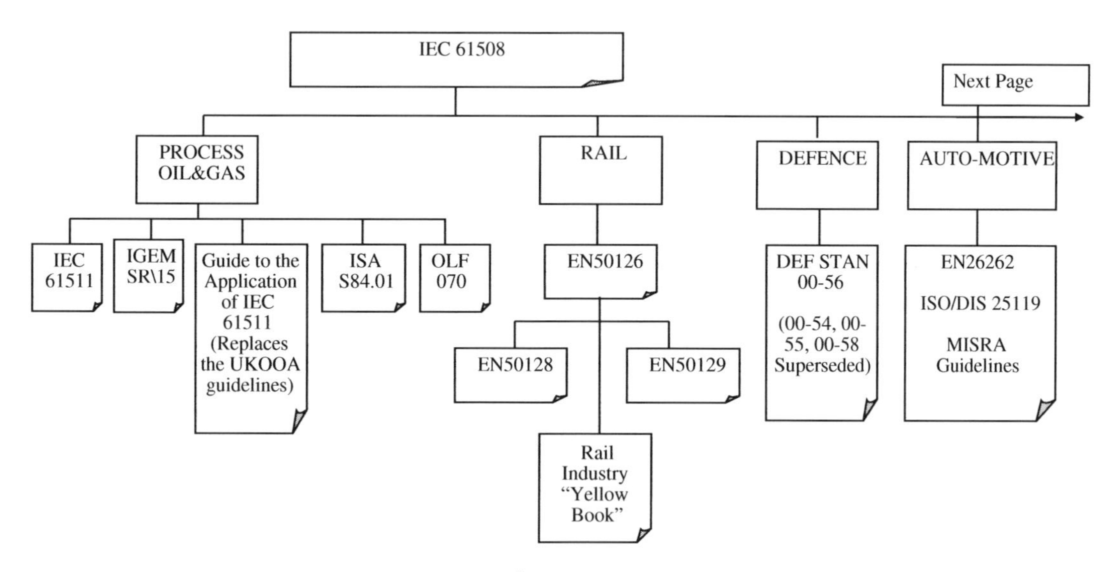

Figure B.1

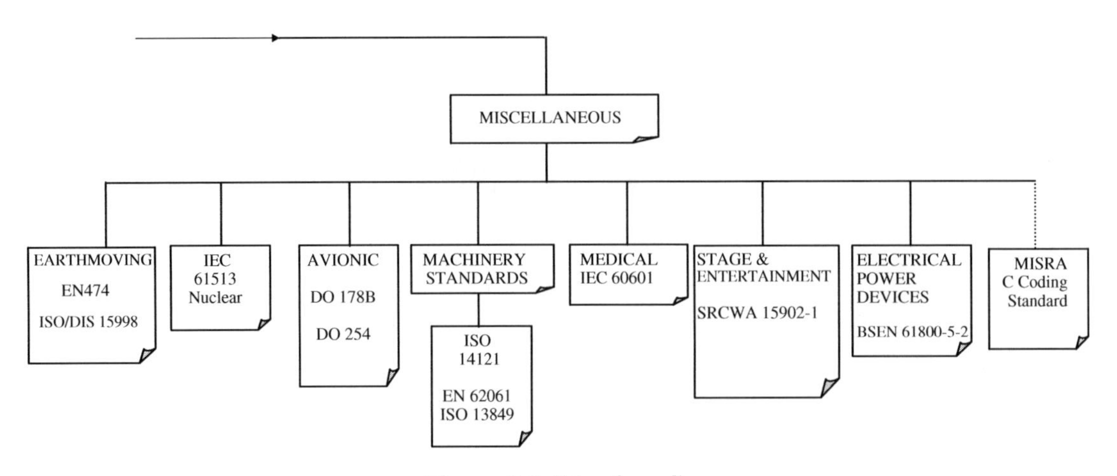

Figure B.1 (Continued)

CHAPTER 8

Second-tier Documents – Process, Oil and Gas Industries

Chapter Outline

Safety Critical Systems Handbook. DOI: 10.1016/B978-0-08-096781-3.10008-2

8.1 IEC International Standard 61511: Functional Safety – Safety Instrumented Systems for the Process Industry Sector

IEC 61511 is intended as the process industry sector implementation of IEC 61508.

It gives application specific guidance on the use of standard products for the use in "safety instrumented" systems using the proven in use justification. The guidance allows the use of field devices to be selected based on being proven in use for application up to SIL 3 and for standard off-the-shelf PLC s for applications up to SIL 2.

The standard was issued at the beginning of 2003 and is in three parts:

Part 1 The normative standard
Part 2 Informative guidance on Part 1
Part 3 Informative guidance on hazard and risk analysis.

Part 1 of the standard covers the life-cycle including

Management of Functional Safety
Hazard and Risk Analysis
Safety instrumented Systems (SIS) Design
through to
SIS decommissioning.

The standard is intended for the activities of SIS system level designers, integrators and users in the process industry.

Suppliers of component-level products, such as field devices and logic solvers, are referred back to IEC 61508 as is everyone in the case of SIL 4.

Part 2 gives general guidance to the use of Part 1 on a paragraph-by-paragraph basis.

Part 3 Gives more detailed guidance on targeting the Safety Integrity Levels and has a number of Appendixes covering both quantitative and qualitative methods.

Since the standard is only aiming at the integration level of the SIS, rather than the individual elements, the requirements for design and development of the SIS (covered by Parts 2 and 3 of IEC 61508) have been significantly simplified. Hardware design has been replaced by a top-level set of straightforward requirements, such as, *"unless otherwise justified the system shall include a manual shutdown mechanism which bypasses the logic solver"*. The software requirements are restricted to the applications software using either limited variability languages or fixed programs. Thus, the software requirement tables that are given in Part 3 of IEC 61508 have been expressed in textual terms using the requirements for SIL 3 but, in general, confined to the "HR" items and using engineering judgment on the suitability at the applications level. For applications software using full variability languages the user is referred to IEC 61508.

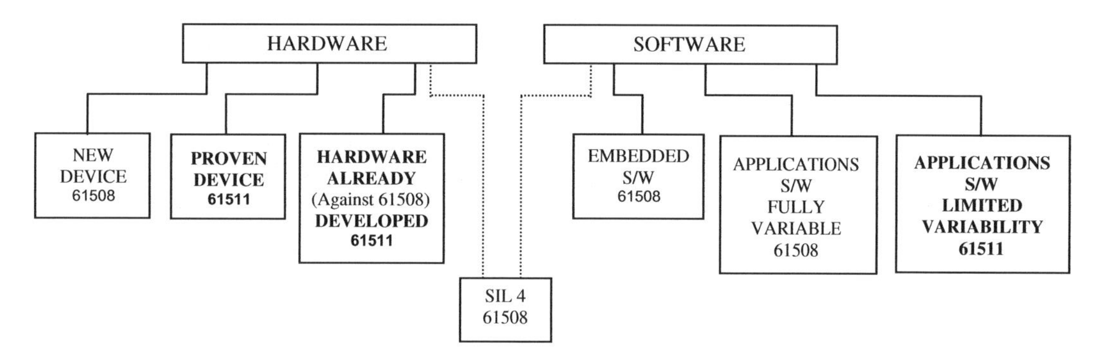

Figure 8.1: IEC 61511 vs IEC 61508.

The techniques and measures detailed within IEC 61511, and hence this chapter, are suitable for the development and modification of the E/E/PE system architecture and software using limited variability languages up to SIL 3 rated safety functions. Unless specifically identified the same techniques and measures will be used for SILs 1, 2 and 3.

Where a project involves the development and modification of a system architecture and application software for SIL 4, or the use of full variability languages for applications software (or the development of a subsystem product), then IEC 61508 should be used.

Figure 8.1 shows the relationship between 61511 and 61508.

8.1.1 Organizing and Managing the Life-cycle

The requirements for the management of functional safety and life-cycle activities are basically the same as given in IEC61508 and are therefore covered by the preceding chapters. The life-cycle is required to be included in the project Quality and Safety Plan.

Each phase of the life-cycle needs to be verified for:

- Adequacy of the outputs from the phase against the requirements stated for that particular phase
- Adequacy of the review, inspection and/or testing coverage of the outputs
- Compatibility between the outputs generated at different life-cycle phases
- Correctness of any data generated
- Performance of the installed safety-related system in terms of both systematic and hardware failures compared to those assumed in the design phase
- Actual demand rate on the safety system compared with the original assessment.

If at any stage of the life-cycle, a change is required which affects an earlier life-cycle phase, then that earlier phase (and the following phases) need to be re-examined and, if changes are required, repeated and re-verified.

The assessment team should include at least one senior, competent person not involved in the project design. All assessments will be identified in the safety plan and, typically, should be done

- After the hazard and risk assessment
- After the design of the safety-related system
- After the installation and development of the operation/maintenance procedures
- After gaining operational/maintenance experience
- After any changes to plant or safety system.

The requirement to perform a hazard and risk analysis is basically the same as for IEC 61508 but with additional guidance being given in Part 3.

Part 1 of 61511 describes the typical layers of risk reduction (namely Control and monitoring, Prevention, Mitigation, Plant emergency response and Community emergency response). All of these should be considered as means of reducing risk and their contributing factors need to be considered in deriving the safety requirement for any safety instrumented system, which form part of the PREVENTION layer.

Part 3 gives examples of numerical approaches, a number of risk graphs and LOPA (as covered in section 2.1.2 of Chapter 2).

8.1.2 Requirements Involving the Specification

The system Functional Design Specification (FDS) will address the PES system architecture and application software requirements. The following need to be included:

- Definition of safety functions, including SIL targets
- Requirements to minimize common cause failures
- Modes of operation, with the assumed demand rate on the system
- A description of process measurements (with their trip points) and output actions
- Subsystem and component selection referencing evidence of suitability at the specified SIL requirement
- Hardware architecture
- Hardware fault tolerance
- Capacity and response time performance that is sufficient to maintain plant safety
- Environmental performance
- Power supply requirements and protection (e.g. under/over voltage) monitoring
- Operator interfaces and their operability including:
    Indication of automatic action
    Indication of overrides / bypasses
    Indication of alarm and fault status
- Procedures for non-steady-state of both the plant and Safety System, i.e. start up, resets etc.

- Action taken on bad process variables (e.g. sensor value out of range, detected open circuit, detected short circuit)
- Software self-monitoring, if not part of the system-level software
- Proof tests and diagnostic test requirements for the logic unit and field devices
- Repair times and action required on detection of a fault to maintain the plant in a safe state
- Identification of any sub-components that need to survive an accident event (e.g. an output valve that needs to survive a fire)
- Design to take into account human capability for both the operator and maintenance staff
- Manual means of independently (to the logic unit) operating the final element should be specified unless otherwise justified by the safety requirements.

Safety functions will be described using semi-formal methods such as cause and effect charts, logic diagrams or sequence charts.

8.1.3 Requirements for Design and Development

(a) Selection of components and subsystems

Components and sub-systems for use in safety instrumented systems should either be in accordance with IEC 61508 or meet the requirements for selection based on prior use given in IEC 61511 as summarized below.

The standard gives guidance on the use of field devices and non-PE logic solvers for up to SIL 3 safety functions using proven-in-use justification and for PE logic solvers, such as standard PLC, guidance on the use for up to SIL 2 safety functions using proven in use justification.

For non-PE logic solvers and field devices (no software, up to SIL 3) the requirements are based on:

- Manufacturer's Quality and Configuration Management
- Adequate identification and specification
- Demonstration of adequate performance in similar operation
- Volume of experience.

For field Devices (FPL software, up to SIL 3) the requirements are based on:

- As above
- Consider I/P and O/P characteristics; mode of use; Function and configuration.

For SIL 3 formal assessment required.

For logic solvers (Up to SIL 2) the requirements are based on:

- As for Field devices
- Experience must consider SIL; complexity; and functionality

- Understand unsafe failure modes
- Use of configurations that address failure modes
- Software has a history in safety-related applications
- Protection against unauthorized/unintended modification
- Formal assessment for SIL 2 applications.

(b) Architecture (i.e. safe failure fraction)

The standard provides two minimum configuration tables, one for the PE logic solvers, the other for non-PE logic solvers and field devices. Both tables are, **unfortunately, formatted differently to the IEC 61508 table** and assume type B sub-systems only (i.e. the typical sub-systems used in the process industry are not assumed to be simple devices and/or do not have good reliability data). For the PE logic solvers the maximum practical SFF is assumed to be between 90% and 99%. For the non-PE logic solvers and field devices a SFF of between 60% and 90% is assumed. The standard actually states that the dominant failure mode is to the safe state or detected, hence this is effectively a relaxation from 60% to 50%. Also the standard gives a list of conditions in the form of proven-in-use and, if a fixed programme device with restricted configurability, then the device can be considered a type A device and hence the required redundancy can be reduced by one. At any time the table in IEC 61508 can be used (see Chapter 3.3.2). The 61511 version is shown below.

PE/LOGIC	SFF < 60%	SFF 60–90%	SFF > 90%	
SIL				
1	1	0	0	
2	2	1	0	Type B
3	3	2	1	
4	See IEC 61508 Part 2 Table 2 (Chapter 3)			
NON PE	SFF < 60 %	SFF 60–90%	SFF > 90%	
SIL				
1	***0***	0		
2	***1***	1		Type B shown thus
3	***2***	2		***Type A (Simple) shown thus***
4	See IEC 61508 Part 2 Table 3 (Chapter 3)			

The 0 represents simplex. The 1 represents m out of m + 1 etc.

(c) Predict the random hardware failures

Random hardware failures will be predicted as already covered in Chapters 5 and 6.

(d) Software

(i) Requirements

The application software architecture needs to be consistent with the hardware architecture and satisfy the safety-integrity requirements.

The application software design shall:

- Be traceable to the requirements
- Be testable
- Include data integrity and reasonableness checks as appropriate
  - Communication link end to end checks (rolling number checks)
  - Range checking on analogue sensor inputs (under and over-range)
  - Bounds checking on data parameters (i.e. have minimum size and complexity).

(ii) Software library modules

Previously developed application software library modules should be used where applicable.

(iii) Software design specification

A software design specification will be provided detailing:

- Software architecture
- The specification for all software modules and a description of connections and interactions
- The order of logical processing
- Any non-safety-related function that is not designed in accordance with this procedure and evidence that it cannot affect correct operation of the safety-related function.

A competent person, as detailed in the Quality and Safety Plan, will approve the software design specification.

(iv) Code

The application code will:

- Conform to an application specific Coding Standard
- Conform to the Safety Manual for the Logic Solver where appropriate
- Be subject to code inspection.

(v) Programming support tools

The standard programming support tools provided by the logic solver manufacturer will be utilized together with the appropriate safety manual.

8.1.4 Integration and Test (Referred to as Verification)

The following minimum verification activities need to be applied:

- Design review on completion of each life-cycle phase
- Individual software module test
- Integrated software module test.

Factory acceptance testing will be carried out to ensure that the logic solver and associated software together satisfy the requirements defined in the safety requirements specifications. This will include:

- Functional test of all safety functions in accordance with the Safety Requirements
- Inputs selected to exercise all specified functional cases
- Input error handling
- Module and system level fault insertion
- System response times including "flood alarm" conditions.

8.1.5 Validation (Meaning Overall Acceptance Test and Close-out of Actions)

System validation will be provided by a factory acceptance test and a close-out audit at the completion of the project.

The complete system shall be validated by inspection and testing that the installed system meets all the requirements, that adequate testing and records have been completed for each stage of the life-cycle and that any deviations have been adequately addressed and closed out. As part of this system validation the application software validation, if applicable, needs to be closed out.

8.1.6 Modifications

Modifications will be carried out using the same techniques and procedures as used in the development of the original code. Change proposals will be positively identified, by the project safety authority, as safety-related or non-safety-related. All safety-related change proposals will involve a design review, including an impact analysis, before approval.

8.1.7 Installation and Commissioning

An installation and commissioning plan will be produced which prepares the system for final system validation. As a minimum the plan should include checking for completeness (earthing, energy sources, instrument calibration, field devices operation, logic solver operation and all operational interfaces). Records of all the testing results shall be kept and any deviations evaluated by a competent person.

8.1.8 Operations and Maintenance

The object of this phase of the life-cycle is to ensure that the required SIL of each safety function is maintained and to ensure that the hazard demand rate on the safety system and the availability of the safety system are consistent with the original design assumptions. If there are any significant increases in hazard demand rate or decreases in the safety system availability between the design assumptions and those found in the operation of the plant which would compromise the plant safety targets then changes to the safety system will have to be made in order to maintain the plant safety.

The operation and maintenance planning need to address

- Routine and abnormal operation activities
- Proof testing and repair maintenance activities
- Procedures, measures and techniques to be used
- Recording of adherence to the procedures
- Recording of all demands on the safety systems along with its conformance to these demands
- Recording of all failures of the safety system
- Competency of all personnel
- Training of all personnel.

8.1.9 Conformance Demonstration Template

In order to justify that the SIL requirements have been correctly selected and satisfied, it is necessary to provide a documented assessment.

The following Conformance Demonstration Template is suggested as a possible format.

Under "Evidence" enter a reference to the project document (e.g. spec, test report, review, calculation) which satisfies that requirement. Under "Feature" take the text in conjunction with the fuller text in this chapter and/or the text in the IEC 61511 Standard.

Activity	Feature (Up to SIL 3 application software)	Evidence
General	Existence of S/W development plan including: procurement, development, integration, verification, validation and modification activities. rev number, configuration management, configured items, deliverables, responsible persons. Evidence of review	
	Clear documentation hierarchy (Q&S Plan, Functional Spec, Design docs, Review strategy, Integration and test plans etc.)	
	Adequate configuration management as per company's FSM procedure	

(Continued)

Activity	Feature (Up to SIL 3 application software)	Evidence
Requirement	A software safety requirements specification including	
	• Revision number, configuration control, author(s) as specified in the Q&S plan	
	• Reviewed, approved, derived from Func Spec	
	• All modes of operation considered, support for FS and nonFS functions clear	
	• External interfaces specified	
	• Baselines and change requests	
	• Clear text and some graphics, use of checklist or structured method, complete, precise, unambiguous and traceable	
	• Describes SR functions and their separation, performance requirements, well defined interfaces, all modes of operation	
Validation Planning	A validation plan explaining technical and procedural steps including: revision number, configuration management, when and who responsible, pass/fail criteria, test environment	
	Plan reviewed	
	Tests have chronological record	
	Records and close-out report	
	Calibration of equipment	
Design and Development	Structured S/W design, recognized methods, under configuration management	
	Use of standards and guidelines	
	Visible and adequate design documentation	
	Modular design with minimum complexity whose decomposition supports testing	
	Readable, testable code (each module reviewed)	
	Small manageable modules (and modules conform to the coding standards)	
	Internal data are not erroneously duplicated and appropriate out-of-range action	
	Structured methods	
	Trusted and verified modules	

(Continued)

Activity	Feature (Up to SIL 3 application software)	Evidence
Language and Support tools	Language fully defined, seen to be error free, unambiguous features, facilitates detection of programming errors, describes unsafe programming features	
	Coding standard/manual (fit for purpose and reviewed)	
	Confidence in tools	
Integration and Test	Overall test strategy in Q&S Plan showing steps to integration and including test environment, tools and provision for remedial action	
	Test specs, reports/results and discrepancy records and remedial action evidence	
	Test logs in chronological order with version referencing	
	Module code review and test (documented)	
	Integration tests with specified test cases, data and pass/fail criteria	
	Pre-defined test cases with boundary values	
	Response times and memory constraints	
	Functional and black box testing	
Modification	Modification log	
	Change control with adequate competence	
	Software configuration management	
	Impact analysis documented	
	Re-verification of affected modules	
Verification	The results of each phase shall be checked to confirm the adequacy of the output against the requirements	
Validation	Validate that each safety function, software and hardware, meets the safety requirements, this is commonly completed as part of the FAT .	

8.2 Institution of Gas Engineers and Managers IGEM/SR/15: Programmable Equipment in Safety-related Applications — 5th Edition 2010

This is the gas industry second-tier guidance to IEC 61508. It is suitable for oil and gas and process applications.

SR/15 describes both quantitative and risk matrix approaches to establishing target SILs but a **very strong** preference for the quantitative approach is stressed. It addresses the setting of maximum tolerable risk targets (fatality rates). The tolerable risk targets were shown in Chapter 2 of this book.

Cost per life saved and ALARP are also addressed.

In order to avoid some of the repetition present in 61508, the life-cycle activities are summarized into three chapters such as provide:

- those common to hardware and software
- those specific to hardware
- those specific to software.

Detailed lists of headings are offered for such essential documents as the Safety Plan, the Safety Specification, the Safety Manual and the Functional Safety assessment.

Some specific design guidance is given for pressure and flow control, gas holder control, burner control, fire and gas detection and process shutdown systems.

There is a worked example of an assessment of a gas detection system.

SR/15 also includes a checklist schedule to aid conformity in the rigor of carrying out assessments based on Appendix 2 of this book. The term "Required" is used to replace the more cumbersome "Highly Recommended" of IEC 61508. The document has 107 pages.

8.3 Guide to the Application of IEC 61511 to Safety Instrumented Systems in the UK Process Industries

This replaces the former UKOOA document: **Guidelines for Process Control and Safety Systems** on Offshore Installations. It was prepared by representatives of EIC, EEMUA, Oil and Gas UK (formerly UKOOA) and HSE and addresses the responsibility and deliverables of organizations involved in the specification, supply, and maintenance of safety instrumented systems.

This guide is applicable to process industries such as onshore and offshore oil and gas, non-nuclear power generation, chemicals and petrochemicals. Other process industries may choose to use the guidelines at their own discretion. It outlines general information for all users plus guidance on organizational responsibilities for end users, designers, suppliers (of systems and products), integrators, installers and maintainers. It does not provide checklists or detail on how to design, operate and maintain such systems.

Clause 3 provides an overview of IEC 61511-1, Clause 4 provides an overview of the legal aspects, Clause 5 focuses on issues that affect all users, and Clause 6 addresses activities of specific users covering the whole life-cycle of the SIS. Technical detail and examples are given in the annexes

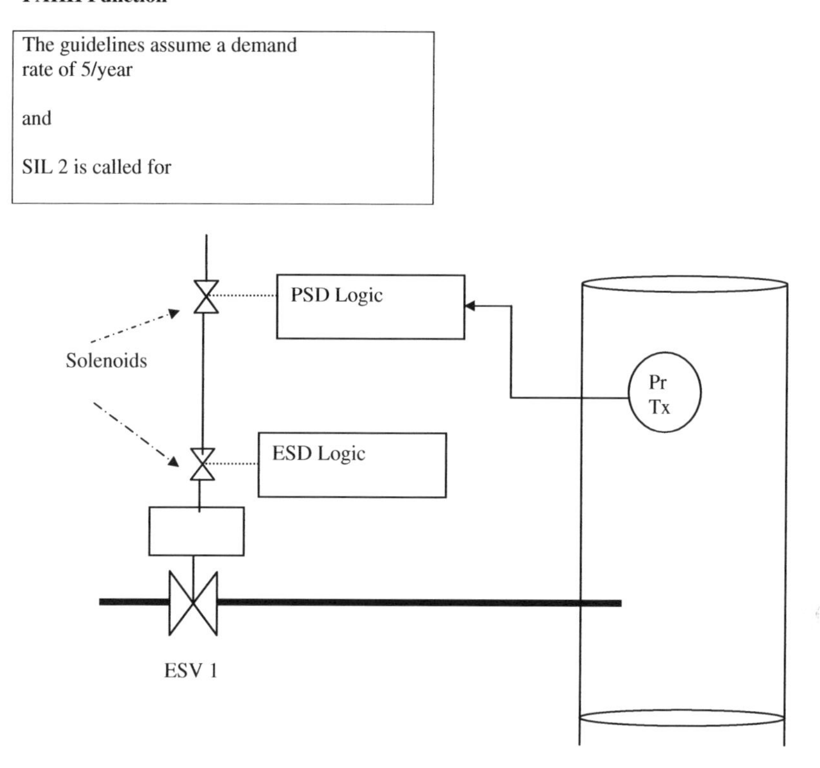

Figure 8.2: OLF-070 — process shutdown functions: PAHH, LAHH, LALL.

8.4 ANSI/ISA-84.00.01 (2004) — Functional Safety, Instrumented Systems for the Process Sector

The original, Instrumentation Systems and Automation Society S84.01, 1996: Application of Safety Instrumented Systems for the Process Industries (see Chapter 10.11g), was from 1996 and pre-dated IRC 61511. ISA have now adopted IEC 61511 and have revised ISA84 using the contents of IEC 61511.

An exception is the "grandfather" clause stating that ISA 84 does not need to be applied to plant which predate 2004.

8.5 Recommended Guidelines for the Application of IEC 61508 and IEC 61511 in the Petroleum Activities on the Norwegian Continental Shelf OLF-070

Published by the Norwegian Oil Industry Association, this 46-page document provides typical safety loops along with the recommended configuration and anticipated SIL. It should be noted that these recommended SILs are typically ONE LEVEL higher than would be expected from the conventional QRA approach described in Chapter 2 of this book.

This is the result of a Norwegian law which states that any new standard associated with safety must IMPROVE on what is currently being achieved. Therefore the authors of OLF-070 assessed the current practices in the Norwegian sector and calculated the expected PFDs for each safety loop and determined which SIL band they fitted.

It should also be noted that the guidelines give failure rate figures for systematic, as well as random hardware, failures.

A typical example of a recommended loop design is shown in Figure 8.2.

CHAPTER 9

Machinery Sector

Chapter Outline

This chapter may seem to describe a different "rule based" graph-type approach to the methods encouraged throughout this book. It has to be said that the authors believe these to be not fully "calibrated" (i.e. dimensioned) against assessments from comparative quantified risk assessment approaches or from field failure data.
However, the methods have stood the test of considerable use and thus represent a benchmark which has become acceptable throughout the sector.

There are two new machinery standards which include, as part of their scope, the area that EN 954 covered. At the time of going to press for the 3rd edition of this book the standard EN 954 has not been made obsolete. The successor standards are EN ISO 13849 and EN 62061.

9.1. EN ISO 14121

EN ISO 14121 (replacing EN 1050) provides guidance on undertaking general risk assessments associated with a machine and, if it is found necessary to provide risk reduction using an active interlock/control mechanism, the evaluation of both the requirements and design of this interlock/control mechanism can be undertaken by using either EN ISO 13849 or EN 62061 as illustrated in Figure 9.1.

EN ISO 14121 provides guidance on the principle of overall risk assessment. It covers all types of risk, not just "functional safety".

Part 1 of the standard provides general guidance on carrying out risk assessments on a machine operation assuming no protective measures. If as the result of this assessment there is a risk, not

Safety Critical Systems Handbook. DOI: 10.1016/B978-0-08-096781-3.10009-4

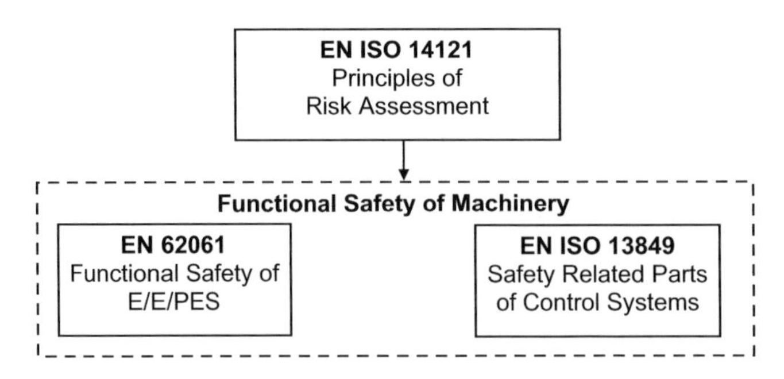

Figure 9.1: Machine safety standards.

considered negligible, then appropriate protective measures need to be applied and the risk assessment repeated to ascertain whether the risk has become negligible. This process is repeated until the risk is negligible, as shown in Figure 9.2.

The risk assessment is required to take into account:

- The risks associated with all phases of a machine's life (i.e. construction, transport, commissioning, assembly and adjustment)
- The intended use of the machine: correct use, non-industrial/domestic use and reasonably foreseeable misuse
- The compatibility of the spatial limits around the machine and its range of movement
- The level of training, ability and experience of the foreseeable users of the machine.

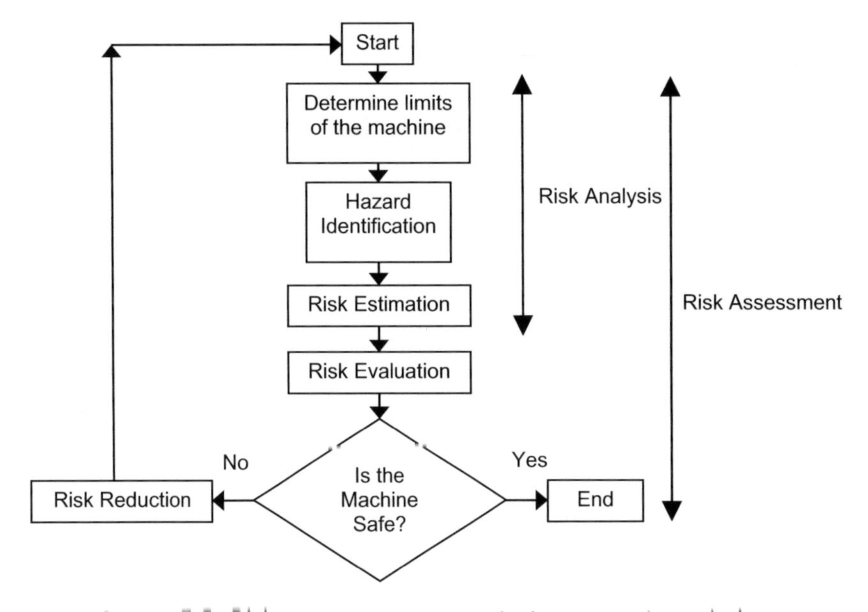

Figure 9.2: Risk assessment approach during machine design.

Consequences	Severity	Class Cl (Fr+Pr+Av)				
		3-4	5-7	8-10	11-13	14-15
Reversible, first aid	1	Negligible Risk	Negligible Risk	Negligible Risk	Low Risk	Low Risk
Reversible, medical attention	2	Negligible Risk	Negligible Risk	Low Risk	Low Risk	Significant Risk
Permanent, losing fingers	3	Negligible Risk	Low Risk	Low Risk	Significant Risk	Significant Risk
Losing an eye or arm	4	Low Risk	Low Risk	Significant Risk	Significant Risk	High Risk
Death	5	Low Risk	Significant Risk	Significant Risk	High Risk	High Risk

Frequency (Fr)	
≤ 1 h	5
> 1 h to ≤ 24 h	4
> 24 h to ≤ 2 w	3
> 2 w to ≤ 1 y	2
> 1 y	1

Probability (Pr)	
Very high	5
Likely	4
Possible	3
Rarely	2
Negligible	1

Avoidance (Av)	
Impossible	5
Possible	3
Likely	1

Figure 9.3: General hazard risk assessment.

The existing risk reduction measures such as guarding, procedures and signage are disregarded when identifying the hazards. When considering the relative merits of different protection measures, any assessment should be weighted to consider (1) as the most effective and (4) as the least effective.

1. Risk reduction by design, i.e. eliminate hazard at the design stage
2. Safeguarding, i.e. safety-related control function (functional safety)
3. Information for use, i.e. signage
4. Additional precautions, i.e. procedures.

Part 2 of the standard provides guidance and examples of methods of undertaking risk estimation. These methods include risk matrix, risk graph, numerical scoring, quantitative methods and hybrid approaches. Figure 9.3 shows the hybrid method from the standard. The object of the assessment is to achieve a 'negligible risk' for all hazards.

If from the general risk assessment some form of "Safety Related Control Function" (SRCF) is required then there is a choice of which of the two standards (EN ISO 13849 or EN 62061) to follow in order to assess the safety requirements for each safety function and how to assess that any proposed system meets the requirements. In general if the safety protection is an electrical-based system either standard could be used. Figure 9.4 gives guidance on which is the more suitable standard based on the type of technology to be used for the safety function.

9.2. EN ISO 13849

This examines complete safety functions, including all the sub-systems included in the design of the safety-related parts of the control system (SRP/CS).

Integrity of SRP/CS and safety function is expressed in terms of performance levels (PL). Control risk assessment is used to determine the **required** PL (PLr) using a risk graph: see Figure 9.5.

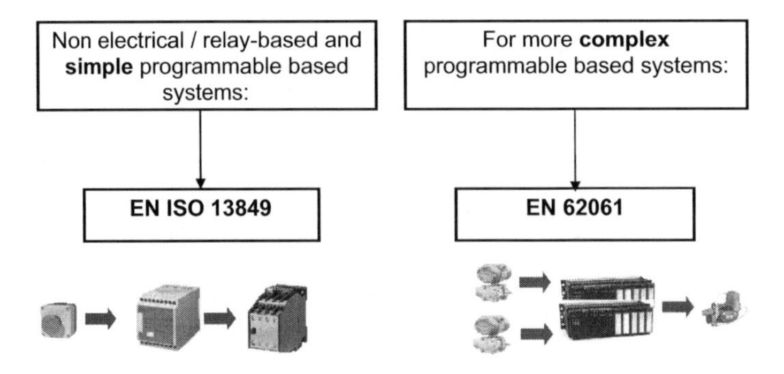

Figure 9.4: Selecting the standard for the design of the SRCF.

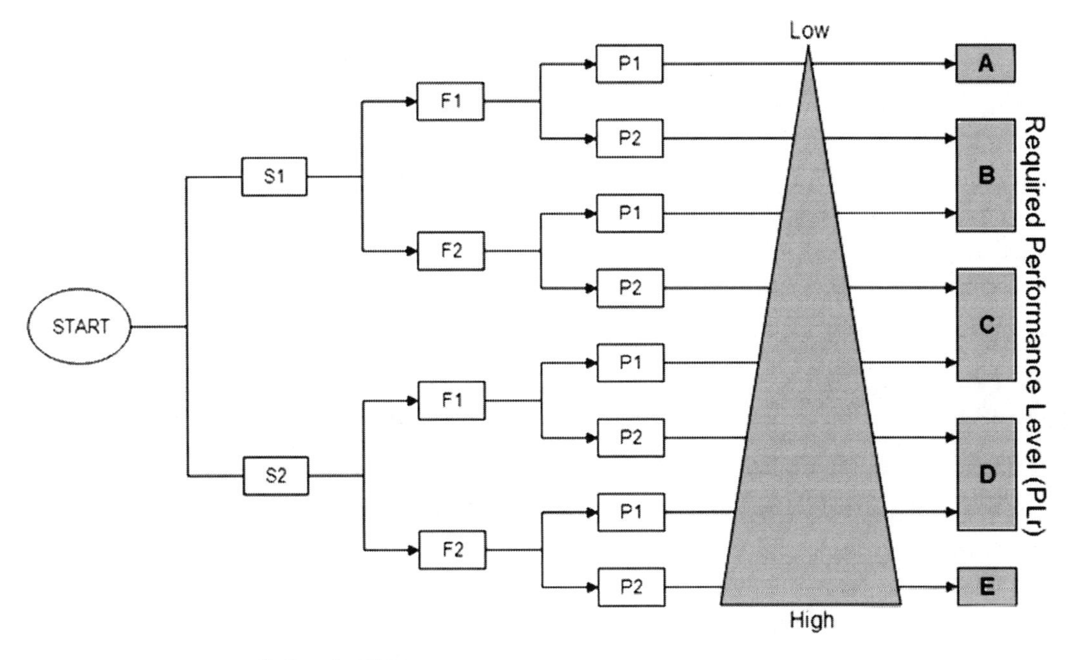

S - Severity of Injury
S1 = Slight (normally reversible injury)
S2 = Serious (normally irreversible) injury including death

F - Frequency and/or exposure time to the hazard
F1 = Seldom to quite often and/or the exposure time is short
F2 = Frequent to continuous and/or the exposure time is long

P - Probability of avoiding the hazard
P1 = Possible under specific conditions
P2 = Scarcely possible

Figure 9.5: Determining the performance level required for each risk.

The design of the SRP/CS and safety function can then be undertaken based on the required level of the PL and the PL Verification of the safety function requires assessment of:

- Diagnostic Coverage (DC)
- Architecture (category)
- Mean Time To Dangerous Failure (MTTFd)
- Common Cause Failures (CCF).

Diagnostic Coverage (DC) is a measure of the effectiveness of diagnostics, expressed as a percentage (DC_{av}) of a safety function, and is calculated from assessing both the total dangerous failure rate and the dangerous detected failure rate for each component in the SRP/CS, and calculating the safety function average DC :

$$DC_{av} = \frac{\sum(\lambda_{DD})}{\sum(\lambda_{D})}$$

DC_{av} then is compared with this table to determine the coverage band:

Coverage	Range of DC
None	$DC < 60\%$
Low	$60\% \leq DC < 90\%$
Medium	$90\% \leq DC < 99\%$
High	$99\% \leq DC$

The Architecture of a safety function is presented in a similar way to IEC 61508 and is shown in Figure 9.6:

However, the architecture is assessed in terms of five categories:

Cat.	Requirements	System behavior
B	• Apply basic safety principles	A fault can cause a loss of the safety function.
	• Can withstand expected influences	
1	• Category B	A fault can cause a loss of the safety function.
	• Well tried components	
	• Well tried safety principles	
2	• Category B	A fault occurring between the checks can cause a loss of the safety function.
	• Well tried safety principles	
	• Functional check at start up and periodically (on/off check)	

(*Continued*)

Cat.	Requirements	System behavior
3	• Category B	Accumulation of undetected faults can cause a loss of the safety function.
	• Well tried safety principles	
	• Single fault does not cause a loss of safety function	
	• Where practicable that fault should be detected	
4	• Category B	Faults will be detected in time to prevent a loss of safety function
	• Well tried safety principles	
	• An accumulation of faults does not cause a loss of safety function	

The architectures are shown in Figures 9.7–9.11.

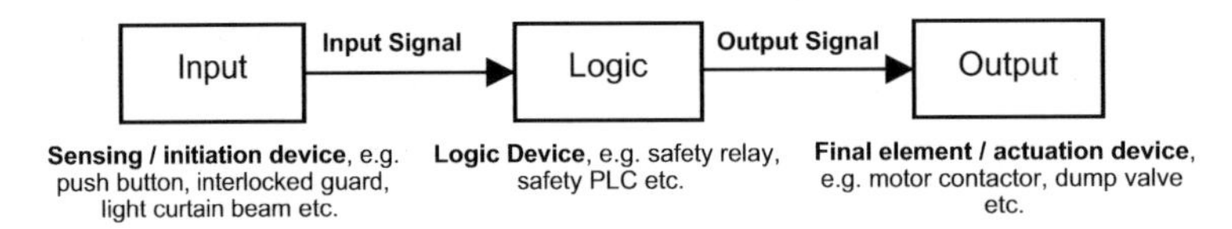

Figure 9.6: Architecture.

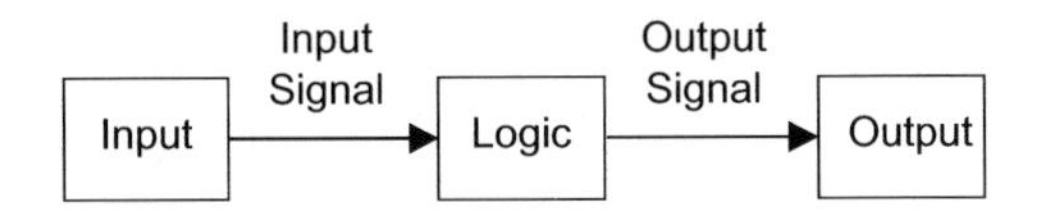

Figure 9.7: Category B architecture.

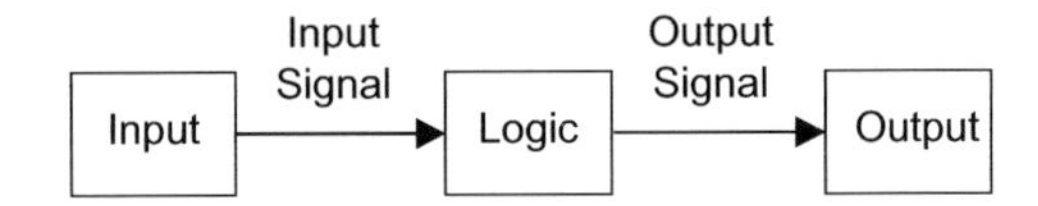

Figure 9.8: Category 1 architecture.

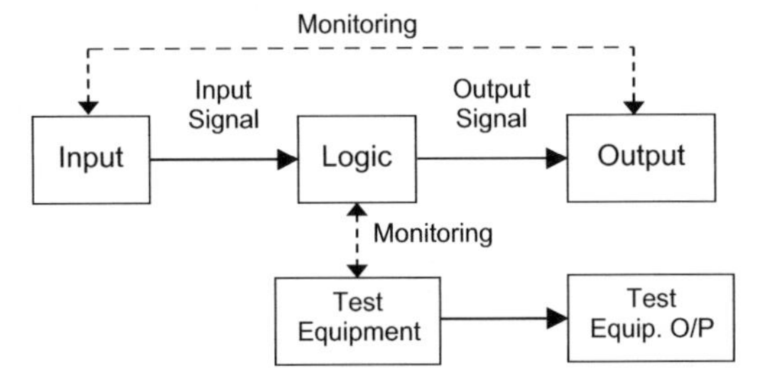

Figure 9.9: Category 2 architecture.

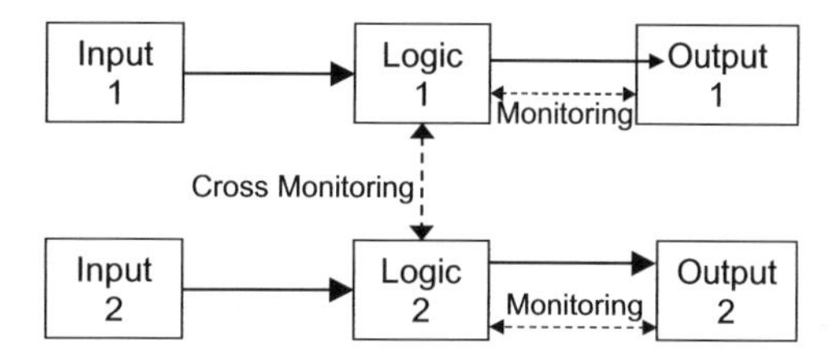

Figure 9.10: Category 3 architecture.

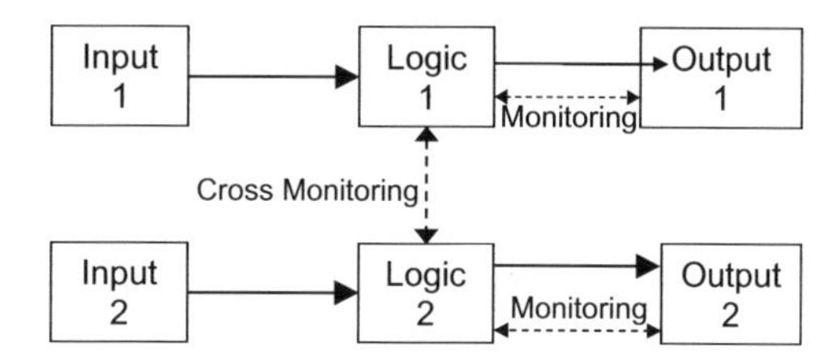

Figure 9.11: Category 4 architecture.

The Assessment

The **mean time to dangerous failure** (MTTFd), includes BOTH the dangerous undetected AND the dangerous detected failures. The total MTTFd of a single safety function channel is calculated from:

$$\text{MTTFd}_{\text{Channel}} + 1/\text{MTTFd}_1 + 1/\text{MTTFd}_2 + 1/\text{MTTFd}_3 + \ldots 1/\text{MTTFd}_n$$

The MTTFd of a channel is then compared with the following table to determine whether the MTTFd is within a given band:

Assessment	Range of MTTFd per channel
Low	3 years <= MTTFd < 10 years
Medium	10 years <= MTTFd < 30 years
High	30 years <= MTTFd < 100 years

The Category, DC_{av} and the MTTFd (per channel) are then compared with the following table in order to determine the performance level (PL) of the SRP/CS and safety function:

Category	B	1	2	2	3	3	4
DC_{av}	None	None	Low	Medium	Low	Medium	High
MTTFd per channel							
Low	a	Not covered	a	b	b	c	Not covered
Medium	b	Not covered	b	c	c	d	Not covered
High	Not covered	c	c	d	d	d	e

In addition, if the design of the safety function includes redundant elements then the Common Cause Failures (CCF) have to be evaluated. The various measures that can affect CCF have to be evaluated, providing a score against each measure. The greater the effectiveness against CCF the higher the score, as shown below. To ensure an adequate design a score of greater than 65 is required.

No.	Measure against CCF	Score
1	Separation/segregation	15
2	Diversity	20
3	Design/application/experience	20
4	Assessment/analysis	5
5	Competence/training	5
6	Environmental	35

9.2.1. *Systematic Failures*

Techniques/procedures/documentation requirements are a very much simplified requirement of that given in IEC 61508 and are more in-line with those given in IEC 61511 (application-level requirements) and consist of;

- Requirement specification for the SRP/CS and safety functions
- Design and integration
- Verification and validation
- Modification
- Documentation.

The design and integration includes requirement for behavior on detection of faults/selection of all components to function within manufacturer's requirements/use of de-energization for the safe state/electromagnetic immunity/clear, modular and documented application software.

9.3. *BS EN 62061*

This is the closest to being the sector specific standard to IEC 61508 and is intended to provide functional safety guidance for the design of safety-related electrical and electronic control systems for machinery and covers the whole life-cycle as covered in IEC 61508.

9.3.1. *Targets*

The integrity of a safety-related electrical control system (SRECS) is expressed using the SIL concept. A risk assessment has to be undertaken to determine the required SIL, typically, using risk matrices as follows.

SIL assignment

Frequency and duration Fr		Probability of hazard event Pr		Avoidance Av	
<= 1 hour	5	Very high	5		
>1 hr - <=1 day	5	Likely	4		
>1 day - <= 2 wks	4	Possible	3	Impossible	5
>2 wks - <= 1 yr	3	Rarely	2	Possible	3
> 1 yr	2	Negligible	1	Likely	1

Consequence	Severity (Se)	Class Cl = Fr + Pr + Av				
		Classes 3–4	Classes 5–7	Classes 8–10	Classes 11–13	Classes 14–15
Death, losing eye or arm	4	SIL2	SIL2	SIL2	SIL3	SIL3
Permanent, losing fingers	3		(AM)	SIL1	SIL2	SIL3
Reversible, medical attention	2			(AM)	SIL1	SIL2
Reversible, first aid	1				(AM)	SIL1

9.3.2. Design

The design of the SRECS can then be undertaken based on the SIL target. SIL verification of the SRECS is very similar to the requirements of IEC 61508 for a continuous/high-demand system:

- Probability of dangerous failure per hour (PFH_D) requirements
- Architecture/Diagnostic Coverage (DC)
- Techniques/procedures/documentation
- Functional safety management.

PFHD requirements are the same as for the IEC 61508 high-demand table (Table 1.1 in Chapter 1) with the exception that SIL 4 is not used in the machinery standards. As in IEC 61508, common cause failures have to be considered when there are redundant paths.

Architecture/Diagnostic Coverage requirements are the same as for IEC 61508, see section 3.3.2 of Chapter 3 for type B components (type A component table is not used), with the exception that SIL 4 is not used in the machinery standards

Techniques/procedures/documentation requirements are a very much simplified version of that given in IEC 61508 and are more in line with those given in IEC 61511 (application-level requirements) and consist of;

- Requirement specification for the SRCFs
- Design and integration
- Verification and validation
- Modification
- Documentation.

The design and integration includes requirement for behavior on detection of faults/selection of all components to function within manufacturer's requirements/use of de-energization for the safe state/electromagnetic immunity/clear, modular and documented application software.

Functional safety management requires that a safety plan is produced to identify the required activities/strategy for SRECs design, application software, integration, verification and validation.

There is a general relationship between PLs and SILs:

Category B	PL a	-
Category 1	PL b	SIL 1
Category 2	PL c	
Category 3	PL d	SIL 2
Category 4	PL e	SIL 3

CHAPTER 10

Other Industry Sectors

Chapter Outline

Safety Critical Systems Handbook. DOI: 10.1016/B978-0-08-096781-3.10010-0

In a book of this type it is impossible to cover all the sector guidance, which, in any case, is expanding rapidly. However, the following are a few of the many documents which now proliferate. They are often referred to as "second tier" guidance in relation to IEC 61508. Due to the open ended nature of the statements made, and to ambiguity of interpretation, it cannot be said that conformance with any one of them automatically implies compliance with IEC 61508.

They tend to cover much the same ground as each other albeit using slightly different terms to describe documents and life-cycle activities.

The figure preceding Chapter 8 illustrates the relationship of the documents to IEC61508. A dotted line indicates that the document addresses similar issues whilst not strictly being viewed as second tier.

10.1 Rail

10.1.1 European Standard EN 50126: Railway Applications – The Specification and Demonstration of Dependability, Reliability, Maintainability and Safety (RAMS)

The development of standards for the design and demonstration of the safety of (in the main) programmable electronic systems for railway-related application has led to the development of a suite of standards. This suite provides both an approach that supports the (general) requirements of IEC61508, and also a means to encourage European rail industry interoperability. This latter element has become increasingly important through the development of Technical Specifications for Interoperability (TSIs) for railway lines classified as suitable for High Speed and Conventional operation. The certification of European railway equipment and systems as "fit for purpose" requires a certification of their "interoperability", that is, their ability to be applied to any member state railway, primarily in order to encourage competition and sustainable growth within the EU member states' railway undertakings

EN 50126 is effectively the European-wide Rail Industry second-tier general guidance (1999) for IEC61508. It is often referred to as "the RAMS standard", as it addresses both reliability and safety issues. EN50126 is intended to cover the railway system in total, while the companion standards, EN 50128 and EN50129, are more specific. CENELEC describe standard 50126 as being "....intended to provide Railway Authorities and the railway support industry throughout the European Community with a process which will enable the implementation of a consistent approach to the management of RAMS".

Risks are assessed by the "risk classification" approach whereby severity, frequency, consequence, etc. are specified by guidewords and an overall "risk classification matrix" obtained. "Intolerable", "ALARP" and "Negligible" categories are thus derived and one proceeds according to the category assessed. The acceptance (or otherwise) of risk is based on choosing a risk acceptance (or hazard tolerability) scheme, the principles of which can be applied throughout the member states (or indeed by other railway authorities). Examples of acceptable risk classifications schemes given include "ALARP" in Great Britain, "GAMAB" (Globalement au moins aussi bon) in France, and "MEM" (Minimum Endogenous Mortality) in Germany. In general terms, the first two schemes deal with global or total risk, whereas the scheme applied in Germany assesses risk to individuals.

The standard is life-cycle based, using the "V-curve" life-cycle approach (i.e. 'V' model). This means that requirements are stated (and subsequently verified and validated) throughout the concept, specification, design and implementation stages of a project. Input and outputs (i.e. deliverables) are described for the life-cycle activities.

10.1.2 EN 50126, EN 50128 and EN 50129

EN 50126 is concerned with the more general specification for the RAMS requirements of a total railway system and the necessary risk assessment, including development of SIL targets and their subsequent satisfactory demonstration, which includes the control of the activities.

CENELEC Standard EN 50128, 2002 "Railway Applications: Software for Railway Control and Protection Systems" covers the requirements for software for railway control and protection systems. It is stated by CENELEC that "The standard specifies procedures and technical requirements for the development of programmable electronic systems for use in railway control and protection applications. The key concept of the standard is the assignment of levels of integrity to software. Techniques and measures for 5 levels of software integrity are detailed".

BS EN 50129, 2002 "Railway Applications, Safety-related Electronics for Signalling" provides details for (hardware and software) for railway control and protection systems. EN 50129 has been produced as a European standardization document defining requirements for the acceptance and approval of safety-related electronic systems in the railway signaling field.

The requirements for safety-related hardware and for the overall system are defined in this standard. It is primarily intended to apply to "fail-safe" and "high integrity" systems such as main line signaling.

The requirements for 50128 and 50129 are those that are most similar (in detail) to the requirements of IEC61508. Thus the suite of three standards provides the overall response to IEC61508, with the three railway-specific documents being roughly equivalent to the Part 1, 2, 3 structure of IEC 61508.

10.1.3 Engineering Safety Management (known as The Yellow Book) – Issue 4.0 2005

This is published by the Rail Safety and Standards Board on behalf of the UK rail industry. It is now at Issue 4.0 and embraces maintenance. The main headings are:

- Engineering Safety Management Fundamentals (Volume 1)
- Obligations and liabilities
- Putting the fundamentals into practice
- Engineering Safety Management Guidance (Volume 2)
- General high-level guidance
- Organization Fundamentals
    - Safety responsibility
    - Organizational goals; Safety culture
    - Competence and training
    - Working with suppliers
    - Communicating safety-related information; Co-ordination
    - Continuing safety management
- Process Fundamentals
    - Safety planning; Systematic processes and good practice
    - Configuration management; Records
    - Independent professional review
- Risk Assessment Fundamentals
    - Identifying hazards; Assessing risk
    - Monitoring risk
- Risk Control Fundamentals
    - Reducing risk; Safety requirements
    - Evidence of safety; Acceptance and approval

Two documents worth mentioning in this brief summary are:

Railway safety case

Any organization which manages infrastructure or operates trains or stations in the UK must currently write a railway safety case and have it accepted before starting operations. The operator must then follow their safety case. Among other things, the operator's railway safety case must describe:

- its safety policy and arrangements for managing safety
- its assessment of the risk
- how it will monitor safety
- how it organizes itself to carry out its safety policy
- how it makes sure that its staff are competent to do safety-related work.

Engineering safety case

An engineering safety case should show that risk has been controlled to an acceptable level. It should also show a systematic approach to managing safety, in order to show that the assessment of the risk is valid. It should consider the effect that the change or product will have on the rest of the railway, including the effect of any changes to operating and maintenance procedures. Similar safety cases are required by CENELEC standards for signaling projects and products and some other projects, and so are commonly produced for these projects across Europe.

Chapter 4 of Volume 2 specifically provides guidance for Maintenance Management.

10.2 UK MOD Documents

Defence Standard 00-56 (Issue 4.0): Hazard Management for Defence Systems

In the past the Ministry of Defence has had its own suite of standards covering much the same ground. However, DEF STAN 00-56 (as Issue 4.0, 2007) supersedes the earlier suite, which are nevertheless summarized in section 10.11 for information.

The Standard, whose scope includes safety-related programmable systems, adopts a "goal based" approach, stating high level requirements for functional safety. It does not prescribe any specific procedures or measures.

A safety case is called for and has to be argued and supported with evidential claims. The structure is:

Part 1: Requirements: this is largely an exhortation to establish safety management, identify hazards and establish a safety case which will reflect risk assessments and the subsequent demonstration of tolerable risks following appropriate risk reduction.
Part 2: Guidance on complying with Part 1: provides more detail on the practices to be adopted to satisfy Part 1. It comprises four volumes:

1. Interpretation of Part 1 – provides more detail as follows:
   - Requirements (e.g. typical deliverables such as safety cases, hazard logs, safety plans, etc.)

- Roles such as safety specialists, independent auditor, etc.
- Safety management including programmes, audit plans, risk criteria, etc.
- The safety case, its function and the report
- Risk management (e.g. HAZID, HAZOP, ALARP)
- Safety requirements
- Interfaces
- Management of changes
- Audits

2. Additional guidance on complex electronics:
   - Deals with the rigor and detail required to verify and validate complex systems
   - Risk management – addressing issues such as HAZID, risk classification and SILs (previously covered in 00-56 Issue 2.0.
3. Software – a successor to 00-55 (below)
4. Electronic hardware – a successor to 00-54 (below).

10.3 Earth Moving Machinery

10.3.1 EN 474: Earth Moving Machinery – Safety

This is in 12 parts which cover:

- General requirements
- Tractors-dozers
- Loaders
- Backhoe-loaders
- Hydraulic excavators
- Dumpers
- Scrapers
- Graders
- Pipe-layers
- Trenchers
- Earth and landfill compactors
- Cable excavators.

Electronic systems are addressed by calling up ISO/DIS 15998.

10.3.2 ISO/DIS 15998: Earth Moving Machinery – MCS using Electronics

This refers to the machine control systems of earth moving vehicles. It calls for requirements to be stated for the foreseen environmental conditions and for a risk analysis to be carried out. Some test criteria are listed as, for example, relative humidities of 30% and 90%, temperatures of −25°C and +70°C with temperature change criteria.

Annexes provide:

- Risk graph approaches for operator and for third-party risks.
- Template systems specification
- List of proven components
- Recommendations for communications bus architectures.

This document also references IEC 61508 as a suitable standard to be met.

10.4 C Coding Standard (MISRA – Motor Industries Research Association) – Development Guidelines for Vehicle Based Programmable Systems)

The MISRA C guidelines were originally intended for the automotive sector but are very well thought of and have been adopted across many industries.

The document provides a subset of the C language for use up to SIL 3. It contains many rules for the use of the language in safety-related applications.

It starts with the premise that the full C language should not be used for safety-related systems. It explains the need for a subset and describes how to use it but, nevertheless, assumes familiarity and competence with the language. It recommends against the use of assembly language in this context.

The contents can be summarized as:

1. Background: covering language insecurities, compiler issues, safety-related uses and standardization
2. Vision: a chapter on the rationale for the subset
3. Developing the subset
4. Scope: covering language issues, applicability, SILs (C++ is excluded) and auto-code
5. Using MISRA C: a chapter on managing and implementing the subset
6. Introduction to the rules: a general introduction
7. Rules: the detailed guidance including character sets, initialization, control flow, pointers, libraries etc.

Further information can be obtained from www.misra.org.uk.

10.5 Automotive

10.5.1 ISO/DIS 26262: Road Vehicles – Functional Safety

This document is the adaptation of IEC 61508 to comply with needs specific to electronic systems within road vehicles. It provides an automotive safety life-cycle (management, development, production, operation, service, decommissioning) and addresses the activities during those life-cycle phases.

There is an automotive-specific risk-based approach for determining risk classes known as “Automotive Safety Integrity Levels, ASILs”.

It lists requirements for validation and confirmation measures to ensure that a sufficient and acceptable level of safety is being achieved. It address the entire development life-cycle (including requirements specification, design, implementation, integration, verification, validation, and configuration).

Part 6 of the document specifically addresses software. Methods defined by the ISO/DIS 26262 standard should be selected according to the “ASIL” (the higher the ASIL, the more rigorous the methods).

10.5.2 MISRA (Motor Industry Software Reliability Association), 2007: Guidelines for Safety Analysis of Vehicle Based Software

These were published as additional guidance to the 1994 document, Development Guidelines for vehicle based software and are aimed at facilitating the meeting of ISO 26262. They introduce the term “controllability” in that vehicle based safety is very much driver orientated. It refers to the “ability of the driver to control the safety of a situation”. The contents cover:

- Safety management
  - Structure, culture, competence, etc.
- Safety process
  - Safety life cycle much as in IEC 61508 (i.e. analyse, plan, realize, validate, etc.)
- Preliminary safety analysis
  - HAZID, risk classification, risk assessment safety plans etc
- Detailed safety analysis
  - Assessment of random hardware failures and defenses against systematic failures.

Appendices include HAZOP, FMEA and fault tree analysis. The document has 98 pages.

The MISRA Risk levels are shown in an Appendix E. In summary they are:

Controllability	Acceptable failure rate	MISRA risk level
Uncontrollable	$< 10^{-5}$ pa	4
Difficult to control	$< 10^{-4}$ pa	3
Debilitating	$< 10^{-3}$ pa	2
Distracting	$< 10^{-2}$ pa	1
Nuisance only	$< 10^{-1}$ pa	0

10.5.3 ISO/DIS 25119: Tractors and Machinery for Agriculture

This takes a similar approach to ISO13849 (see Chapter 9).

10.6 IEC International Standard 61513: Nuclear Power Plants – Instrumentation and Control for Systems Important to Safety – General Requirements for Systems

Many of the existing standards that were applicable to the nuclear sector prior to the emergence of IEC 61508 generally adopted a similar approach to IEC 61508. These existing standards are either from IEC or IAEA. Thus the nuclear sector standard IEC 61513 primarily links these existing standards to IEC 61508. The IEC existing standards are **60880, 60987, 61226 and 60964,** and the existing IAEA standards are primarily **NS-R-1, 50-SG-D1, 50-SG-D3** and **50-SG-D8.**

These standards present a similar overall safety cycle and system life-cycle approach as in IEC 61508 with more in-depth details at each stage compared to IEC 61508. **IEC 60964** covers the identification of the required safety function applicable to power plants and **IEC 61226** provides system categorization for different types of safety functions. The SIS design is then covered by **IEC 60987** for hardware design and **IEC 60880** for software design. **IAEA 50-C-D now NS-R-1** covers the overall Safety Design, **50-SG-D1** gives the Classification of Safety Functions, **50-SG-D3** covers all Protection Systems and **50-SG-D8** provides the requirements for the Instrumentation and Control Systems

The current standards do not directly use the SAFETY INTEGRITY LEVELS as given in IEC61508. The standards use the existing categorization (IEC 61226) A, B or C. These are related to 'Safety Functions', A = highest and C = lowest. IEC 61513 adds corresponding system classes, 1 = highest and 3 = lowest, where;

Class 1 system can be used for Cat A, B or C
Class 2 system can be used for Cat B or C
Class 3 system can be used for Cat C.

Categorization A is for safety functions, which play a principal role in maintenance of NPP safety
Categorization B is for safety functions that provide a complementary role to category A
Categorization C is for safety functions that have an indirect role in maintenance of NPP safety.

No specific reliability/availability targets are set against each of these categories or classes. There is, however, a maximum limit set for software based systems of 10^{-4} PFD. More generally the reliability/availability targets are set in the Plant Safety Design Base and can be set either quantitatively or qualitatively. There is a preference for quantitative plus basic requirements on layers and types of protection.

Class 1 / Categorization A is generally accepted as being equivalent to SIL3
Class 2 / Categorization B is generally accepted as being equivalent to SIL2
Class 3 / Categorization C is generally accepted as being equivalent to SIL1.

Architectural constraints do not have a direct relationship with the tables in IEC61508 Part 2, but are summarized as:

CAT A: shall have redundancy, to be fault tolerant to one failure, with separation. Levels of self-test are also given.
CAT B: redundancy is preferred but Simplex system with adequate reliability is acceptable, again levels of self test given.
CAT C: redundancy not required. Reliability needs to be adequate, self-test required.

General design requirements: within this standard and the related standard there is significantly more guidance on each of the steps in the design. In particular:

- Human factors
- Defenses against common cause failures
- Separation/segregation
- Diversity.

There are mapping tables for relating its clauses to the clause numbers in IEC 61508.

10.7 Avionics

10.7.1 RTCA DO-178B/(EUROCAE ED-12B): Software Considerations in Airborne Systems and Equipment Certification

This is a very detailed and thorough standard which is used in civil avionics to provide a basis for certifying software used in aircraft. Drafted by a EUROCAE/RTCA committee, DO-178B was published in 1992 and replaces an earlier version published in 1985. The qualification of software tools, diverse software, formal methods and user-modified software are now included.

It defines five levels of software criticality from A (software which can lead to catastrophic failure) to E (no effect). The Standard provides guidance which applies to levels A to D.

The detailed listing of techniques covers:

Systems aspects: including the criticality levels, architecture considerations, user modifiable software
The software life-cycle
Software planning

Development: including requirements, design, coding and integration
Verification: including reviews, test and test environments
Configuration management: including baselines, traceability, changes, archive and retrieval
Software quality
Certification
Life-cycle data: describes the data requirements at the various stages in the life-cycle.

Each of the software quality processes/techniques described in the Standard is then listed (10 pages) and the degree to which they are required is indicated for each of the criticality levels A to D. The mapping is:

Level	SIL
A	4
B	3
C	2
D	1
E	Not safety-related

10.7.2 RTCA/DO-254: Design Assurance Guidance for Airborne Electronic Hardware

This is a counterpart to the above DO-178B, being launched in 2005. It specifically addresses complex electronic hardware and includes FPGAs (field programmable gate arrays) and ASICs (application specific integrated circuits). The same levels A-E apply (see DO-178B). The main sections include:

- System aspects of hardware design
- Hardware design life-cycle
- Planning
- Design processes
- Validation and verification
- Configuration management
- Certification.

Previously developed hardware is addressed, along with commercial off-the-shelf components.

10.8 Medical — IEC 60601: Medical Electrical Equipment, General Requirements for Basic Safety and Essential Performance

The Standard requires manufacturers of electro-medical equipment to have a formal risk management system in place. Manufacturers must estimate the risks relating to their device and take action dependent upon how that risk compares to predefined levels of acceptability.

There are objective pass/fail criteria and one may choose simply to follow such requirements in the design of their device.

The risk management process must be documented, like a quality management system, and the manufacturer must establish acceptable risks for a device, based upon regulations, standards, state-of-the-art and other relevant factors.

IEC 60601 addresses four basic areas:

- *Mechanical* – is the equipment enclosure strong enough to endure the wear and tear of normal use? Are moving parts properly protected to ensure a safety hazard is not created? Is the unit stable and lacking sharp corners, edges, etc.?
- *Markings* –the Standard defines a list of data that must be present on the product's name-plate including information on its electrical requirements, together with a test protocol for the durability of markings.
- *Earthing* – this defines how the device is attached to the earth or safety ground connection of an electrical power supply to provide safety in the event of an electrical fault.
- *Electrical* – addresses electrical safety as it relates to the process of caring for the patient. The standard requires that the system operate safely in the event of a "single fault" condition.

10.9 Stage and Theatrical Equipment

10.9.1 SR CWA 15902-1:2009 Lifting and Load-bearing Equipment for Stages and other Production Areas within the Entertainment Industry

This document covers all machinery used in the entertainment industry including machinery that is excluded from the Machinery Directive and gives a significant amount of prescriptive guidance on a range of safety aspects for the mechanical parts of the system and refers to EN 60204-32 associated with the electrical design and IEC 61508 with regard to the use of programmable electronic systems. Currently it is common practice for control systems, such as controllers of winches for use in flying objects on a stage which could lead to harm to the actors, to be verified as meeting SIL 3.

Typical applications include but are not limited to the following:

- acoustic doors
- auditorium and compensating elevators
- cycloramas
- fire curtains
- fly bar systems (manual, motor driven)
- lighting bars
- movable lighting towers and stage platforms

- movable proscenium arches
- orchestra elevators
- performer flying
- point hoists
- projection screens (manual or motor-driven)
- revolving stages and turntables
- scenery storage elevators
- side stage and rear stage shutters
- stage elevators and wagons
- tiltable stage floors
- trap elevators.

There is (Annex A) a very comprehensive risk assessment list (aid memoir) covering such headings as radiation, noise, thermal hazards, vibration, etc.

10.10 Electrical Power Drives

10.10.1 BS EN 61800-5-2:2007 Adjustable Speed Electrical Power Drive Systems

This standard covers the requirements for functional safety for power drive systems (PDS(SR)) and covers very closely the requirements of IEC 61508 but is limited to up to SIL 3 continuous / high demand applications.

10.11 Documents which are now Withdrawn

(a) UKOOA: Guidelines for Process Control and Safety Systems on Offshore Installations

Replaced by Guide to the Application of IEC 61511 to safety instrumented systems in the UK process industries – see Chapter 8.

(b) EEMUA Guidelines, Publication No 160: Safety-related Instrument Systems for the Process Industry (Including Programmable Electronic Systems)

These were published, in 1989, by EEMUA (Engineering Equipment and Materials Users Association) in response to the HSE PES guidance. They were produced well before the emergence of IEC 61508 drafts.

(c) IEE Publication, SEMSPLC, 1996: Safety-related Application Software for Programmable Logic Controllers

This document was an interpretation, at the time, of the draft 61508 Standard. It provided guidance specific to programmable logic controllers.

(d) MOD Standard 00-54: Requirements for Safety-related Electronic Hardware in Defense Equipment

This complemented 00-55 and 00-56 by covering the hardware aspects. It is life-cycle based and covers much the same ground as IEC 61508 Part 2. The guidance is tailored in rigor according to the SIL. In MOD terms this document is "*cancelled*", which means it is no longer in use and cannot be used in contracts.

(e) MOD Standard 00-55: The Procurement of Safety Critical Software in Defense Equipment

This is akin to Part 3 of IEC 61508 and has superseded the old MOD 00-16 guide to achievement of quality in software. It is far more stringent and is perhaps one of the most demanding standards in this area. In MOD terms this document is "*obsolescent*", which means it will not be updated. It could be invoked in a contract until such time as it is canceled.

Whereas the majority of the documents described here are for guidance, 00-55 is a standard and is intended to be mandatory on suppliers of "safety-critical" software to the MOD. It is unlikely that the majority of suppliers are capable of responding to all of its requirements but the intention was that, over a period of time, industry evolved to adopt it in full.

It dealt with software rather than the whole system and its major requirements include:

- The non-use of assembler language
- The use of static analysis
- A preference for formal methods
- The use and approval of a safety plan
- The use of a software quality plan
- The use of a validation plan
- An independent safety auditor.

(f) MOD Standard 00-58: A Guideline for HAZOP Studies on Systems which Include Programmable Electronic Systems

As the title suggests, this standard describes the HAZOP process in the context of identifying potentially hazardous variations from the design intent. Part 1 is the requirements and Part 2 provides more detailed guidance on such items as HAZOP guidewords for particular types of system, team roles, recording the study, etc. In MOD terms this document is "*cancelled*", which means it is no longer in use and cannot be used in contracts.

(g) Instrumentation Systems and Automation Society S84.01, 1996: Application of Safety Instrumented Systems for the Process Industries

The Instrumentation Systems and Automation Society (USA) is an International Society for measurement and control. They developed S84 as a response to IEC 61508 and it was intended as applications-specific second-tier guidance. It adopted the E/E/PES mnemonic in respect of safety instrumented systems (SIS), namely the sensors, logic solving and final elements in much the same way as IEC 61511.

A life-cycle approach was adopted from process design, through procurement and installation and including operations, maintenance, modifications and de-commissioning. The process starts with a Safety Requirements Specification and moves through the life-cycle with requirements similar to IEC 61508.

An Annex provided detailed design guidance on issues such as sensor diversity, communications, embedded software and electro-mechanical devices. For example, the guidance on sensor diversity suggests:

SIL 1: Single sensor likely to be suitable
SIL 2: Redundancy (identical) with separation
SIL 3: Redundancy (diverse) with separation.

It is replaced by ANSI/ISA-84.00.01 (2004) – see Chapter 8.4.

PART C

Case Studies in the Form of Exercises and Examples

Chapter 11 is an exercise involving SIL targeting for a pressure let-down system. The design is then compared with the target and improvements are evaluated and subjected to ALARP criteria. The answers are provided in Appendix 5.

Chapter 12 is a typical assessment report on a burner control system. The reader can compare and critique this, having read the earlier chapters of this book.

Chapter 13 presents a number of rather different SIL targeting examples.

Chapter 14 is a purely hypothetical proposal for a rail train braking system.

Chapter 15 summarizes some Technis work on helicopter statistics and risk assessments.

Chapter 16 contains case studies relating to tidal gates.

These case studies address the four quantitative aspects of IEC61508:

SIL Targeting
Random Hardware Failures
Safe Failure Fraction
ALARP.

CHAPTER 11

Pressure Control System (Exercise)

Chapter Outline

This exercise is based on a real scenario. Spaces have been left for the reader to attempt the calculations. The answers are provided in Appendix 5.

11.1 The Unprotected System

Consider a plant supplying gas to offsite via a twin stream pressure control station. Each stream is regulated by two valves (top of Figure 11.1). Each valve is under the control of its downstream pressure. Each valve is closed by the upstream gas pressure via its pilot valve, J, but only when its pilot valve, K1, is closed. Opening pilot valve K1 relieves the pressure on the diaphragm of valve V, allowing it to open. Assume that a HAZOP (HAZard and OPerability) study of this system establishes that downstream overpressure, whereby the valves fail to control the downstream pressure, is an event which could lead to one or more fatalities.

Since the risk is offsite, and a two-fatality scenario asssumed, a target maximum tolerable risk of 10^{-5} per annum has been proposed.

Assume that a quantified risk assessment has predicted a probability of 20% that failure, involving overpressure, will lead to subsequent pipe rupture and ignition. Furthermore it is predicted that, due to the high population density, fatality is 50% likely.

Safety Critical Systems Handbook. DOI: 10.1016/B978-0-08-096781-3.10011-2

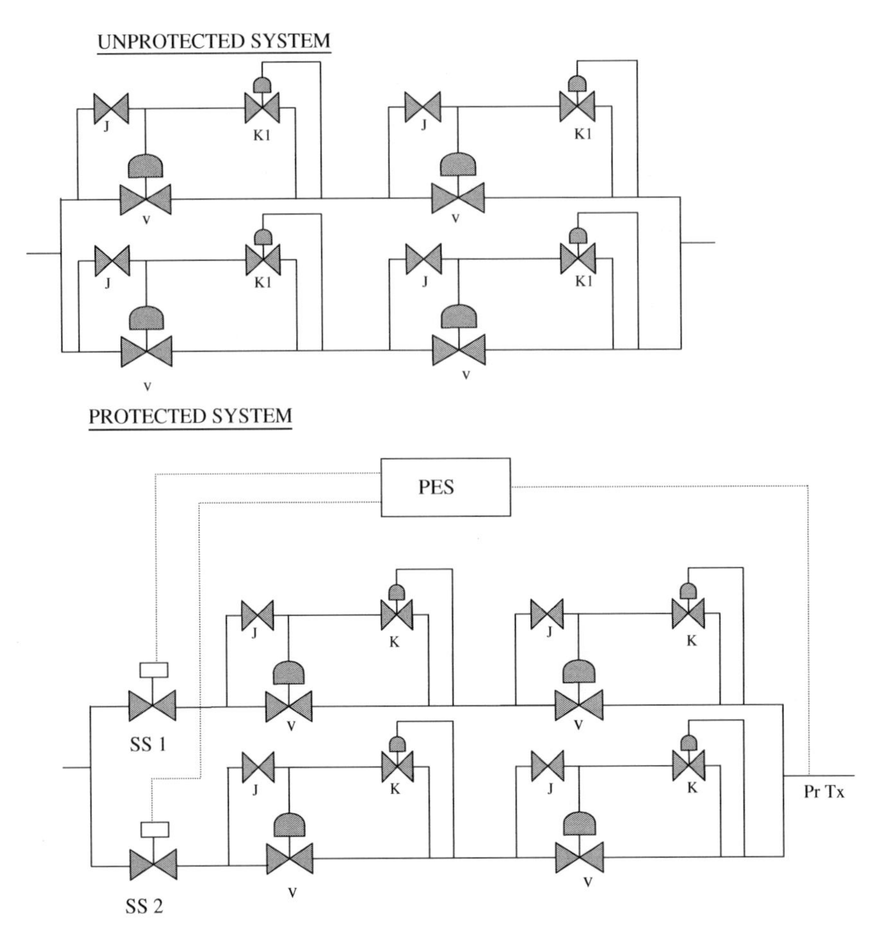

Figure 11.1: The system, with and without backup protection.

Assume also that the plant offers approximately 10 risks in total to the same population.

It follows that the target failure rate for overpressure of the twin stream sub-system is

$$[10^{-5}/[10 \text{ risks} \times 0.2 \times 0.5] = \mathbf{10^{-5}\ pa}].$$

Assume, however, that field experience of a significant number of these twin stream systems shows that the frequency of overpressure is dominated by the pilots and is **2.5 × 10^{-3} pa.**

11.2 Protection System

Since 2.5×10^{-3} is greater than 10^{-5} a design modification is proposed whereby a programmable electronic system (PES) closes a valve in each stream, based on an independent measure of the downstream pressure. The valves consist of actuated ball valves (sprung to close). This is illustrated at the bottom of Figure 11.1.

The target Unavailability for this "add-on" safety system is therefore ....?

which indicates a SIL of ....?

11.3 Assumptions

The following assumptions are made in order to construct and quantify the reliability model:

(a) Failure rates (symbol λ), for the purpose of this prediction, are assumed to be constant with time. Both early and wearout-related failures are assumed to be removed by burn-in and preventive replacement respectively.
(b) The MTTR (mean time to repair) of a revealed failure is 4 hours.
(c) The auto-test coverage of the PLC is 90% and occurs at just under 5 minute intervals. The MDT (mean down time) for failures revealed by this PES auto-test are taken to be the same as the MTTR (mean time to repair) because the MTTR >> the auto-test period. The MDT is thus assumed to be 4 hours. Neither the pressure transmitter nor the valve is assumed to have any self diagnostics.
(d) The manual proof-test is assumed to be 100% effective and to occur annually (ca 8000 hours).
(e) One maintenance crew is assumed to be available for each of the three equipment types (PES, Instrumentation, Pneumatics).
(f) The detailed design assumptions needed for an assessment of the common cause failure BETA factor (see modified proposal) are summarized in section 11.8.

11.4 Reliability Block Diagram

Figure 11.2 is the reliability block diagram for the add-on safety system. Note that the PES will occur twice in the diagram. This is because the model needs to address those failures revealed by auto-test separately from those revealed by the longer manual proof-test due to their different MDTs (explained more fully in Chapter 6.3).

11.5 Failure Rate Data

The following failure rate data will have been chosen for the protection system components, shown in Figure 11.1. These are the component level failure modes which lead to the hazard under consideration (i.e. downstream overpressure). FARADIP.THREE has been used to obtain the failure rates.

TO BE FILLED IN BY THE READER (see Appendix 5 for answer)

Figure 11.2: Reliability block diagram.

Item	Failure mode	Failure rates 10^{-6} per hour	
		Total	Mode
PES	PES low or zero*	5	0.25
Pressure transmitter	Fail low	2	0.5 (25% has been assumed)
Actuated ball valve (sprung to close)	Fail to close	8	0.8**

*This represents any failure of the PES i/p, CPU or o/p causing the low condition.
**10% has been used based on the fact that the most likely failure mode is fail closed.

11.6 Quantifying the Model

The following Unavailability calculations address each of the groups (left to right) in Figure 11.2 (see Appendix 5):

(a) Ball valve 1 - unrevealed failures
Unavailability = ...
= ....

(b) Ball valve 2 - unrevealed failures
Unavailability = ...
= ....

(c) PES output 1 failures revealed by auto-test
Unavailability = ...
= ....

(d) PES output 1 failures not revealed by auto-test
Unavailability = ...
= ....

(e) PES output 2 failures revealed by auto-test
Unavailability = ...
= ....

(f) PES 2 output failures not revealed by auto-test
Unavailability = ...
= ....

(g) Pressure Transmitter - unrevealed failures
Unavailability = ...
= ....
The predicted Unavailability is obtained from the sum of the unavailabilities in (a) to (e)
= ....?

11.7 Proposed Design and Maintenance Modifications

The proposed system is not acceptable (as can be seen in Appendix 5) and modifications are required.

Before making modification proposals it is helpful to examine the relative contributions to system failure of the various elements in Figure 11.2.

??% from items (a) and (b) Ball Valve.
??% from items (c) to (f) the PES.
??% from item (g) the Pressure Transmitter

It was decided to duplicate the Pressure Transmitter and vote the pair (1 out of 2). It was also decided to reduce the proof test interval to 6 months (c4000 hrs).

11.8 Modeling Common Cause Failure (Pressure Transmitters)

The BETAPLUS method provides a method for assessing the percentage of common cause failures. The scoring for the method was carried out assuming:

- Written procedures for system operation and maintenance are evident but not extensive
- There is some training of all staff in CCF awareness
- Extensive environmental testing was conducted
- Identical (i.e. non-diverse) redundancy

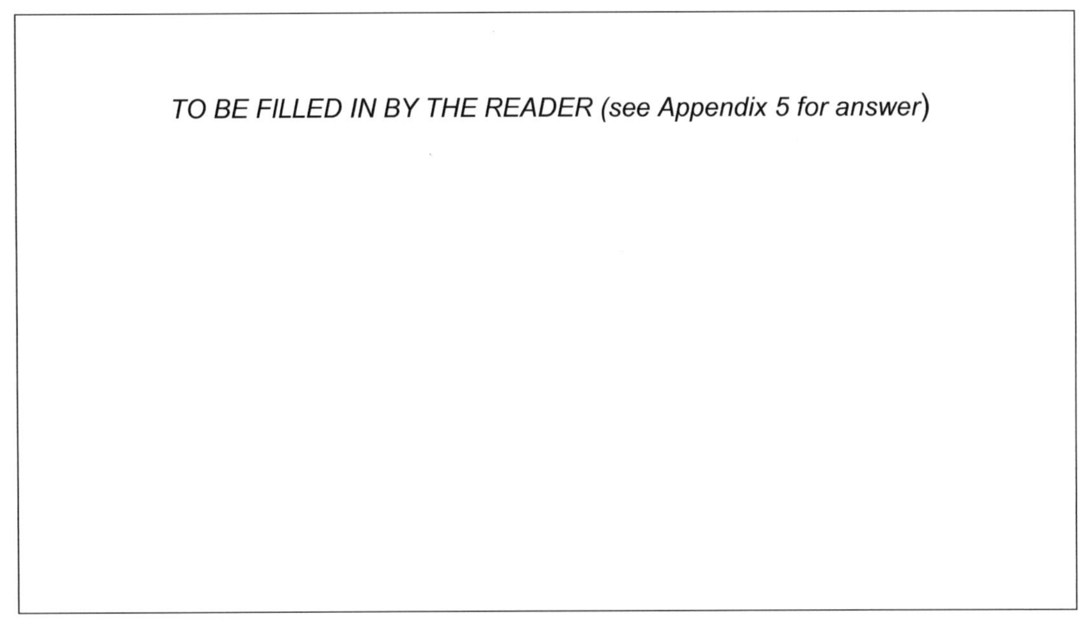

Figure 11.3: Revised reliability block diagram (or fault tree).

- Basic top level FMEA (failure mode analysis) had been carried out
- There is some limited field failure data collection
- Simple, well proven, pressure transmitters ½ metre apart with cables routed together
- Good electrical protection
- Annual proof test.

The BETAPLUS software package performs the calculations and was used to calculate a BETA value of 9%.

11.9 Quantifying the Revised Model

The following takes account of the pressure transmitter redundancy, common cause failure and the revised proof test interval. Changed figures are shown in bold in Appendix 5.

Changed figures are shown in bold.

(a) Ball valve SS1 fails open.
Unavailability = ...
= ....

(b) Ball valve SS2 fails open.
Unavailability = ...
= ....

(c) PES output 1 fails to close valve (Undiagnosed Failure).
Unavailability = ...
= ....

(d) PES output 2 fails to close valve (Undiagnosed Failure).
Unavailability = ...
= ....

(e) PES output 1 fails to close valve (Diagnosed Failure).
Unavailability = ...
= ....

(f) PES output 2 fails to close valve (Diagnosed Failure).
Unavailability = ...
= ....

(g) Voted pair of pressure transmitters.
Unavailability = ...
= ....

(h) Common cause failure of pressure transmitters.
Unavailability = ...
= ....

The predicted Unavailability is obtained from the sum of the unavailabilities in (a) to (h)
= ....?

11.10 ALARP

Assume that further improvements in CCF can be achieved for a total cost of £1,000. Assume, also, that this results in an improvement in unavailability to $\mathbf{4 \times 10^{-4}}$. It is necessary to consider, applying the ALARP principle, whether this improvement should be implemented.

The cost per life saved over a 40 year life of the equipment (without cost discounting) is calculated, assuming two fatalities, as follows:

??????? (see Appendix 5)

11.11 Architectural Constraints

Consider the architectural constraints imposed by IEC 61508 Part 2, outlined in Chapter 3.3.2.

Do the pressure transmitters and valves in the proposed system, meet the minimum architectural constraints assuming they are "TYPE A components"?

Does the PES, in the proposed system, meet the minimum architectural constraints assuming it is a "TYPE B component"?

CHAPTER 12

Burner Control Assessment (Example)

Chapter Outline

Safety Critical Systems Handbook. DOI: 10.1016/B978-0-08-096781-3.10012-4

This chapter consists of a possible report of an integrity study on a proposed replacement burner control system. Unlike Chapter 11, the requirement involves the high demand table and the target is expressed as a failure rate.

This is not intended as a MODEL report but an example of a typical approach. The reader may care to study it in the light of this book and attempt to list omissions and to suggest improvements.

SAFETY INTEGRITY STUDY OF A PROPOSED REPLACEMENT BOILER CONTROLLER

Executive Summary and Recommendations

Objectives

To establish a Safety-Integrity Level target, *vis-à-vis* IEC 61508, for a Boiler Control System which is regarded as safety-related. To address the following failure mode: Pilots are extinguished but nevertheless burner gas continues to be released with subsequent explosion of the unignited gas. To assess the design against the above target and to make recommendations.

Targets

A Maximum Tolerable Risk target of 10^{-4} per annum which leads to a MAXIMUM TOLERABLE TARGET FAILURE RATE of 3×10^{-3} per annum (see Section 12.2).

This implies a SIL 2 target.

Results

The frequency of the top event is 2×10^{-4} pa and the target is met. This result remains within the ALARP region but it was shown that further risk reduction is unlikely to be justified.

Recommendations

Review all the assumptions in Sections 12.2, 12.3 and 12.4.3. Review the failure rates and down times in Section 12.5 and the fault tree logic, in Figures 12.1–12.3, for a future version of this study.

Continue to address ALARP.

Place a SIL 2 requirement on the system vendor, in respect of the requirements of Parts 2 and 3 of IEC 61508.

Because very coarse assumptions have had to be made, concerning the PLC and SAM (safety monitor) design, carry out a more detailed analysis with the chosen vendor.

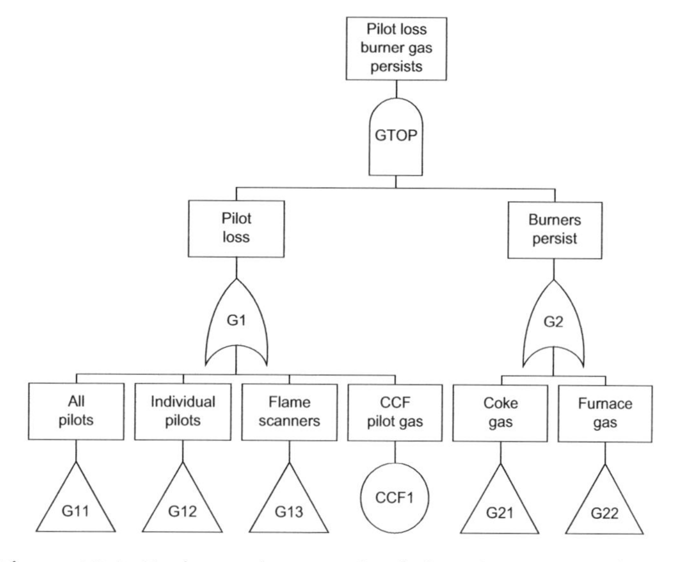

Figure 12.1: Fault tree (suppressing below Gates G1 and G2).

Address the following design considerations with the vendor:

- Effect of loss of power supply, particularly where it is to only some of the equipment.
- Examine the detail of the PLC/SAM interconnections to the I/O and ensure that the fault tree logic is not compromised.
- Establish if the effect of failure of the valve limit switches needs to be included in the fault tree logic.

12.1 Objectives

(a) To establish a Safety-Integrity Level target, *vis-à-vis* IEC 61508, for a Boiler Control System which is regarded as safety-related.
(b) To address the following failure mode: Pilots are extinguished but nevertheless burner gas continues to be released with subsequent explosion of the unignited gas.
(c) To assess the design against the above target.
(d) To make recommendations.

12.2 Integrity Requirements

IGEM SR/15 suggests target maximum tolerable risk criteria. These are, for individual risk:

1–2 FATALITIES (EMPLOYEE)	10^{-4} pa
BROADLY ACCEPTABLE	10^{-6} pa

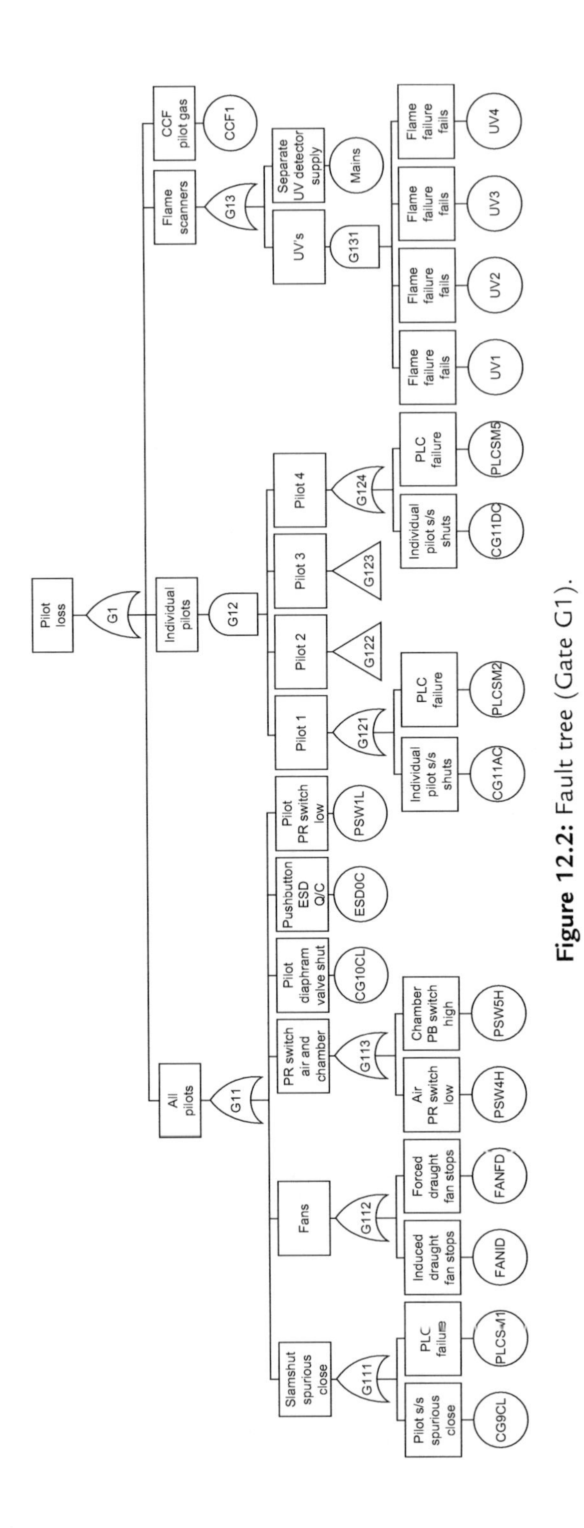

Figure 12.2: Fault tree (Gate G1).

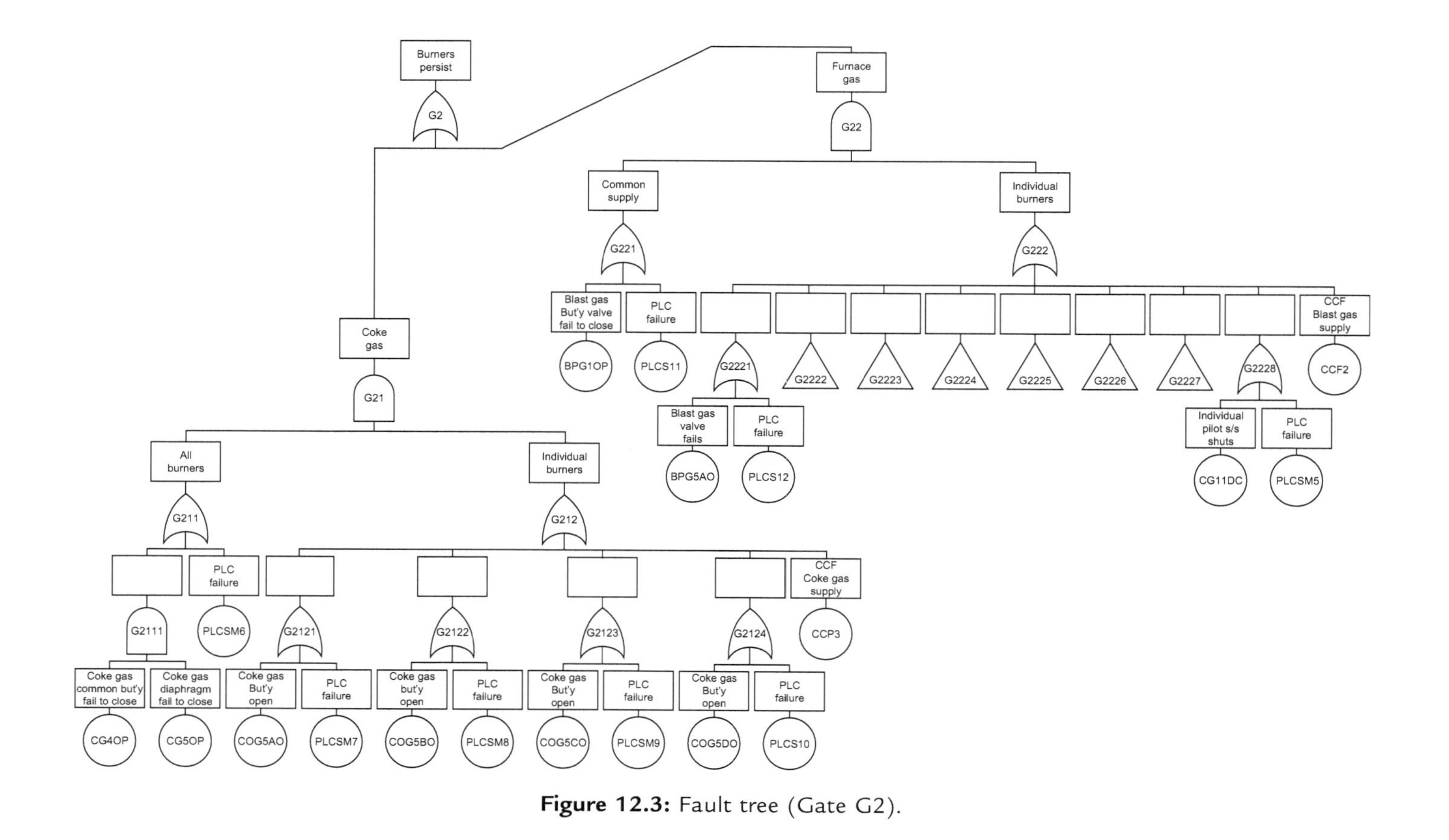

Figure 12.3: Fault tree (Gate G2).

Assume that there is a 0.9 probability of ignition of the unburnt gases.
Assume that there is a 0.1 probability of the explosion leading to fatality.
Assume that there is a 0.5 probability that the oil burners are not active.
Assume that there is a 0.75 probability of there being a person at risk.

Hence the MAXIMUM TOLERABLE TARGET FAILURE
RATE $= 10^{-4}$ pa divided by $(0.9 \times 0.1 \times 0.5 \times 0.75)$
$= 3 \times 10^{-3}$ per annum

This invokes a SIL 2 target.

12.3 Assumptions

12.3.1 Specific

(a) Proof test is carried out annually.Thus the mean down time of unrevealed failures, being half the proof-test interval, is approximately 4000 hours.
(b) The system is in operation 365 days per annum.
(c) The burner control system comprises a combination of four 'XYZ Ltd' PLCs and a number of safety monitors (known as SAMs).

12.3.2 General

(a) Reliability assessment is a statistical process for applying historical failure data to proposed designs and configurations. It therefore provides a credible target/ estimate of the likely reliability of equipment assuming manufacturing, design and operating conditions identical to those under which the data were collected. It is a valuable design review technique for comparing alternative designs, establishing order of magnitude performance targets and evaluating the potential effects of design changes.
(b) Failure rates (symbol), for the purpose of this prediction, are assumed to be constant with time. Both early and wearout related failures would decrease the reliability but are assumed to be removed by burn-in and preventive replacement respectively.
(c) Each single component failure which causes system failure is described as a SERIES ELEMENT. This is represented, in fault tree notation, as an OR gate whereby any failure causes the top event. The system failure rate contribution from this source is obtained from the sum of the individual failure rates.
(d) Where coincident failures are needed to fail for the relevant system failure mode to occur then this is represented, in fault tree notation, as an AND gate where more than one failure is needed to cause the top event.
(e) The failure rates used, and thus the predicted MTBFs (mean time between failure) and availabilities, are those credibly associated with a well proven design after a suitable

period of reliability growth. They might therefore be considered optimistic as far as field trial or early build states are concerned.

(f) Calendar based failure rates have been used in this study.

(g) Software failures are systematic and, as such, are not random. They are not quantified in this study.

12.4 Results

12.4.1 Random Hardware Failures

The fault tree logic was constructed from a discussion of the failure scenarios at the meeting on 8 January 2001 involving Messrs 'Q' and 'Z'. The fault tree was analysed using the TECHNIS fault tree package TTREE.

The frequency of the top event (Figure 12.1) is 2×10^{-4}pa (see Annex 1) which is well within the target.

Annex 1 shows the combinations of failures (cut sets) which lead to the failure mode in question. It is useful to note that at least three coincident events are required to lead to the top event. An 'Importance' measure is provided for each cut set and it can be seen that no cut set contributes more than 1.4% of the total. There is therefore no suggestion of a critical component.

12.4.2 Qualitative Requirements

The qualitative measures required to limit software failures are listed, for each SIL, in the IGEM SR/15 and IEC 61508 documents. Although the IGEM guidance harmonises closely with IEC 61508, compliance with SR/15 does not automatically imply compliance with IEC 61508.

It has to be stressed that this type of qualitative assessment merely establishes a measure of 'adherence to a process' and does not signify that the quantitative SIL is automatically achieved by those activities. It addresses, however, a set of measures deemed to be appropriate (at the SIL) by the above documents.

It should also be kept in mind that an assessment is in respect of the specific failure mode. The assessment of these qualitative measures should therefore, ideally, be in respect of their application to those failure modes rather than in a general sense.

The purpose of the following is to provide an aide-memoire *whereby features of the design cycle can be assessed in greater detail for inclusion in a later assessment. this list is based on safety integrity level (SIL 2).*

1 Requirements

(a) Requirements Definition: This needs to be identified. It needs to be under configuration control with adequate document identification. It should also refer to the safety integrity requirements of the failure mode addressed in this report. *Subject to this, the requirement will be met.*A tender document, in response to the Requirements Specification, might well have been produced by the supplier and might well be identified.
(b) The Functional Specification needs to address the safety integrity requirement and to be specific about the failure modes. It will be desirable to state to the client that it is understood that the integrity issue is 'loss of pilot followed by …' etc. *Subject to this, the requirement will be met.*
(c) The design may not utilize a CAD specification tool or formal method in delineating the requirement. However, the safety-related system might comprise simple control loops and therefore not involve parameter calculation, branching decision algorithms or complex data manipulation. Thus, a formal specification language may not be applicable. The documentation might be controlled by ISO 9001 configuration control and appropriate software management. The need for an additional CAD specification tool may not be considered necessary. *Subject to this, the requirement will be met.*

2 Design and language

(a) There should be evidence of a 'structured' design method. Examples include:
    Logic diagrams
    Data dictionary
    Data flow diagrams
    Truth tables
  Subject to this, the requirement will be met.
(b) There should be a company specific, or better still, project specific coding/design standard which addresses, for example (list where possible):
    Use of a suitable language
    Compiler requirements
    Hygienic use of the language set
    Use of templates (i.e. field proven) modules
    No dynamic objects
    No dynamic variables or online checking thereof
    Limited interrupts, pointers and recursion
    No unconditional jumps
    Fully defined module interfaces
  Subject to this, the requirement will be met.
(c) Ascertain if the compiler/translator certified or internally validated by long use. *Subject to this, the requirement will be met*

(d) Demonstrate a modular approach to the structure of the code and rules for modules (i.e. single entry/exit). *Subject to this, the requirement will be met.*

3 Fault tolerance

(a) Assuming Type B components, and a non-redundant configuration, at least 90% safe failure fraction is required for SIL 2. It will be necessary to establish that 90% of PLC failures are either detected by the watch-dog or result in failures not invoking the failure mode addressed in this study. *Subject to a review the requirement will be met.*
(b) Desirable features (not necessarily essential) would be, error detection/correction codes and failure assertion programming. *Subject to this, the requirement will be met.*
(c) Demonstrate graceful degradation in the design philosophy. *Subject to this, the requirement will be met.*

4 Documentation and change control

(a) A description is needed here to cover: Rigour of configuration control (i.e. document master index, change control register, change notes, change procedure, requirements matrix (customer spec/FDS/ FAT mapping)). *Subject to this, the requirement will be met.*
(b) The change/modification process should be fairly rigorous, key words are:
    Impact analysis of each change
    Re-verification of changed and affected modules (the full test not just the perceived change)
    Re-verification of the whole system for each change
    Data recording during these re-tests
  Subject to this, the requirement will be met.

5 Design review

(a) Formal design review procedure? Evidence that design reviews are:
    Specifically planned in a Quality Plan document
    Which items in the design cycle are to be reviewed (i.e. FDS, acceptance test results etc.)
    Described in terms of who is participating, what is being reviewed, what documents etc.
    Followed by remedial action
    Specifically addressing the above failure mode
    Code review see (b)
  Subject to this, the requirement will be met.
(b) Code: Specific code review at pseudo code or ladder or language level which addresses the above failure mode. *Subject to this, the requirement will be met.*
(c) There needs to be justification that the language is not suitable for static analysis and that the code walkthrough is sufficiently rigorous for a simple PLC language set in that it is a form of 'low level static analysis'. *Subject to this, the requirement will be met.*

6 Test (applies to both hardware and software)

(a) There should be a comprehensive set of functional and interface test procedures which address the above failure mode. The test procedures need to evidence some sort of formal test case development for the software (i.e. formally addressing the execution possibilities, boundary values and extremes). *Subject to this, the requirement will be met.*
(b) There should be mis-use testing in the context of failing due to some scenario of I/O or operator interface. *Subject to this, the requirement will be met.*
(c) There should be evidence of formal recording and review of all test results including remedial action (probably via the configuration and change procedures). *Subject to this, the requirement will be met.*
(d) There should be specific final validation test plan for proving the safety-related feature. This could be during commissioning. *Subject to this, the requirement will be met.*

7 Integrity assessment

Reliability modelling has been used in the integrity assessment.

8 Quality, safety and management

(a) In respect of the safety integrity issues (i.e. for the above failure mode) some evidence of specific competency mapping to show that individuals have been chosen for tasks with the requirements in view (e.g. safety testing, integrity assessment). The competency requirements of IEC 61508 infer that appropriate job descriptions and training records for operating and maintenance staff are in place. *Subject to this, the requirement will be met.*
(b) Show that an ISO 9001 quality system is in operation, if not actually certified. *Subject to this, the requirement will be met.*
(c) Show evidence of safety management in the sense of ascertaining safety engineering requirements in a project as is the case in this project. This study needs to address the safety management system (known as functional safety capability) of the equipment designer and operator. Conformance with IEC 61508 involves this aspect of the safety-related equipment. *Subject to this, the requirement will be met.*
(d) Failure recording, particularly where long term evidence of a component (e.g. the compiler or the PLC hardware) can be demonstrated is beneficial. *Subject to this, the requirement will be met.*

9 Installation and commissioning

There needs to be a full commissioning test. Also, modifications will need to be subject to control and records will need to be kept. *Subject to this, the requirement will be met.*

12.4.3 ALARP

The ALARP (as low as reasonably practicable) principle involves deciding if the cost and time of any proposed risk reduction is, or is not, grossly disproportionate to the safety benefit gained.

The demonstration of ALARP is supported by calculating the Cost per Life Saved of the proposal. The process is described in Chapter 3. Successive improvements are considered in this fashion until the cost becomes disproportionate. The target of 3×10^{-3} pa corresponded to a maximum tolerable risk target of 10^{-4} pa. The resulting 2×10^{-4} pa corresponds to a risk of 6.6×10^{-6} pa. This individual risk is not as small as the BROADLY ACCEPTABLE level and ALARP should be considered.

Assuming, for the sake of argument, that the scenario is sufficiently serious as to involve two fatalities then any proposed further risk reduction would need to be assessed against the ALARP principle. Assuming a £2 000 000 per life saved criterion then the following would apply to a proposed risk reduction, from 6.6×10^{-6} pa. Assuming a 30-year plant life:

$$£2\,000\,000 = \frac{(\text{Proposed expenditure})}{([6.6 \times 10^{-6} - 10^{-6}] \times 30 \times 2)}$$

Thus: proposed expenditure = £672

It seems unlikely that the degree of further risk reduction referred to could be achieved within £672 and thus it might be argued that ALARP is satisfied.

12.5 Failure Rate Data

In this study the FARADIP.THREE Version 6.5 data ranges have been used for some of the items. The data are expressed as ranges. In general the lower figure in the range, used in a prediction, is likely to yield an assessment of the credible design objective reliability. That is the reliability which might reasonably be targeted after some field experience and a realistic reliability growth programme. The initial (field trial or prototype) reliability might well be an order of magnitude less than this figure. The centre column figure (in the FARADIP software package) indicates a failure rate which is more frequently indicated by the various sources. It has been used where available. The higher figure will probably include a high proportion of maintenance revealed defects and failures. F3 refers to FARADIP.THREE, Judge refers to judgement.

Code (Description)	Mode	Failure rate PMH (or fixed per hour probability)	Mode rate 10^{-6}	MDT (hrs)	Reference
CCF1 (Common Cause Failures)	any	0.1	0.1	24	JUDGE
CCF2/3 (Common Cause Failures)	any	0.1	0.1	4000	JUDGE
ESDOC (ESD button)	o/c	0.1	0.1	24	F3
UV (UV detector)	fail	5	2	24	F3
MAINS (UV separate supply)	fail	5	5	24	JUDGE
PLC... (Revealed failures)	–	5	1	24	JUDGE
PLC... (Unrevealed failures)	–	5	1	4000	JUDGE
FAN (Any fan)	fail	10	10	24	F3

(Continued)

Code (Description)	Mode	Failure rate PMH (or fixed per hour probability)	Mode rate 10^{-6}	MDT (hrs)	Reference
PSWL (Pressure switch)	low	2	1	24	F3
PSWH (Pressure switch)	high	2	1	24	F3
CG10CL (Pilot diaphragm vlv)	closed	2	1	24	F3
CG9CL (Slamshut)	sp close	–	1	24	F3
CG11... (Slamshuts)	sp close	–	4	24	F3
COG5... (Butterfly vlv)	fail to close	–	2	4000	F3
CG4OP... (Butterfly vlv)	fail to close	–	2	4000	F3
CG5OP (Diaphragm vlv)	fail to close	–	2	4000	F3
BFG... (Blast gas vlvs)	–	–	2	4000	F3

12.6 References

A reference section would normally be included.

Annex I Fault tree details

File name: Burner.TRO

Results of fault tree quantification for top event: GTOP

Top event frequency	= 0.222E – 07 per hour = 0.194E – 03 per year
Top event MTBF	= 0.451E + 08 hours = 0.515E + 04 years
Top event probability	= 0.526E – 06

Basic event reliability data

Basic event	Type	Failure rate	Mean fault duration
CCF1	I/E	0.100E –06	24.0
CG10CL	I/E	0.100E – 05	24.0
ESDOC	I/E	0.100E – 06	24.0
PSW1L	I/E	0.100E – 05	24.0
CG9CL	I/E	0.100E – 05	24.0
PLCSM1	I/E	0.100E – 05	24.0
FANID	I/E	0.100E – 04	24.0
FANFD	I/E	0.100E – 04	24.0
PSW4H	I/E	0.100E – 05	24.0
PSW5H	I/E	0.100E – 05	24.0

CG11AC	I/E	0.400E − 05	24.0
PLCSM2	I/E	0.100E − 05	24.0
CG11BC	I/E	0.400E − 05	24.0
PLCSM3	I/E	0.100E − 05	24.0
CG11CC	I/E	0.400E − 05	24.0
PLCSM4	I/E	0.100E − 05	24.0
CG11DC	I/E	0.400E − 05	24.0
PLCSM5	I/E	0.100E − 05	24.0
MAINS	I/E	0.500E − 05	24.0
UV1	I/E	0.200E − 05	24.0
UV2	I/E	0.200E − 05	24.0
UV3	I/E	0.200E − 05	24.0
UV4	I/E	0.200E − 05	24.0
PLCSM6	I/E	0.100E − 05	0.400E + 04
CCF3	I/E	0.100E − 06	0.400E + 04
COG5AO	I/E	0.200E − 05	0.400E + 04
PLCSM7	I/E	0.100E − 05	0.400E + 04
COG5BO	I/E	0.200E − 05	0.400E + 04
PLCSM8	I/E	0.100E − 05	0.400E + 04
COG5CO	I/E	0.200E − 05	0.400E + 04
PLCSM9	I/E	0.100E − 05	0.400E + 04
COG5DO	I/E	0.200E − 05	0.400E + 04
PLCS10	I/E	0.100E − 05	0.400E + 04
CG4OP	I/E	0.200E − 05	0.400E + 04
CG5OP	I/E	0.200E − 05	0.400E + 04
BFG1OP	I/E	0.100E − 05	0.400E + 04
PLCS11	I/E	0.100E − 05	0.400E + 04
CCF2	I/E	0.100E − 06	0.400E + 04
BFG5AO	I/E	0.100E − 05	0.400E + 04
PLCS12	I/E	0.100E − 05	0.400E + 04
BFG5BO	I/E	0.100E − 05	0.400E + 04
PLCS13	I/E	0.100E − 05	0.400E + 04
BFG5CO	I/E	0.100E − 05	0.400E + 04
PLCS14	I/E	0.100E − 05	0.400E + 04
BFG5DO	I/E	0.100E − 05	0.400E + 04
PLCS15	I/E	0.100E − 05	0.400E + 04
BFG5EO	I/E	0.100E − 05	0.400E + 04
PLCS16	I/E	0.100E − 05	0.400E + 04
BFG5FO	I/E	0.100E − 05	0.400E + 04
PLCS17	I/E	0.100E − 05	0.400E + 04
BFG5GO	I/E	0.100E − 05	0.400E + 04
PLCS18	I/E	0.100E − 05	0.400E + 04
BFG5HO	I/E	0.100E − 05	0.400E + 04
PLCS19	I/E	0.100E − 05	0.400E + 04

Barlow-Proschan measure of cut set importance (Note: This is the name given to the practice of ranking cut sets by frequency)

Rank 1 Importance 0.144E − 01 MTBF hours 0.313E + 10 MTBF years 0.357E + 06

Basic event	Type	Failure rate	Mean fault duration
FANID	I/E	0.100E − 04	24.0
PLCSM6	I/E	0.100E − 05	0.400E + 04
COG5AO	I/E	0.200E − 05	0.400E + 04

Rank 2 Importance 0.144E − 01 MTBF hours 0.313E + 10 MTBF years 0.357E + 06

Basic event	Type	Failure rate	Mean fault duration
FANID	I/E	0.100E − 04	24.0
PLCSM6	I/E	0.100E − 05	0.400E + 04
COG5BO	I/E	0.200E − 05	0.400E + 04

Rank 3 Importance 0.144E − 01 MTBF hours 0.313E + 10 MTBF years 0.357E + 06

Basic event	Type	Failure rate	Mean fault duration
FANID	I/E	0.100E − 04	24.0
PLCSM6	I/E	0.100E − 05	0.400E + 04
COG5CO	I/E	0.200E − 05	0.400E + 04

Rank 4 Importance 0.144E − 01 MTBF hours 0.313E + 10 MTBF years 0.357E + 06

Basic event	Type	Failure rate	Mean fault duration
FANID	I/E	0.100E − 04	24.0
PLCSM6	I/E	0.100E − 05	0.400E + 04
COG5DO	I/E	0.200E − 05	0.400E + 04

Rank 5 Importance 0.144E − 01 MTBF hours 0.313E + 10 MTBF years 0.357E + 06

Basic even	Type	Failure rate	Mean fault duration
FANFD	I/E	0.100E −04	24.0
PLCSM6	I/E	0.100E − 05	0.400E + 04
COG5AO	I/E	0.200E − 05	0.400E + 04

Rank 6 Importance 0.144E − 01 MTBF hours 0.313E + 10 MTBF years 0.357E + 06

Basic event	Type	Failure rate	Mean fault duration
FANFD	I/E	0.100E − 04	24.0
PLCSM6	I/E	0.100E − 05	0.400E + 04
COG5BO	I/E	0.200E − 05	0.400E + 04

CHAPTER 13

SIL Targeting – Some Practical Examples

Chapter Outline

13.1 A Problem Involving EUC/SRS Independence

Figure 13.1 shows the same EUC as was used in Chapter 11. In this case, however, the additional protection is provided by means of additional K2 pilot valves, provided for each valve, V. This implies that failure of the valves, V, was (wrongly) not perceived to be significant. Closing the K2 pilot valve (via the PES and an I/P converter) has the same effect as closing the K1 pilot. The valve, 'V', is thus closed by either K1 or K2. This additional safety-related protection system (consisting of PES, I/P converters and K2 pilots) provides a backup means of closing valve 'V'.

The PES receives a pressure signal from the pressure transmitters P. A "high" signal will cause the PES to close the K2 pilots and thus valves 'V'.

It might be argued that the integrity target for the add-on SRS (consisting of PESs, transmitters and pilots) is assessed as in Chapter 11. This would lead to the same SIL target as is argued in Chapter 11, namely **2.5×10^{-3} PFD being SIL 2**.

However, there are two reasons why the SRS is far from INDEPENDENT of the EUC:

(a) Failures of the Valve V actuators, causing the valves to fail open, will not be mitigated by the K2 pilot.

Safety Critical Systems Handbook. DOI: 10.1016/B978-0-08-096781-3.10013-6

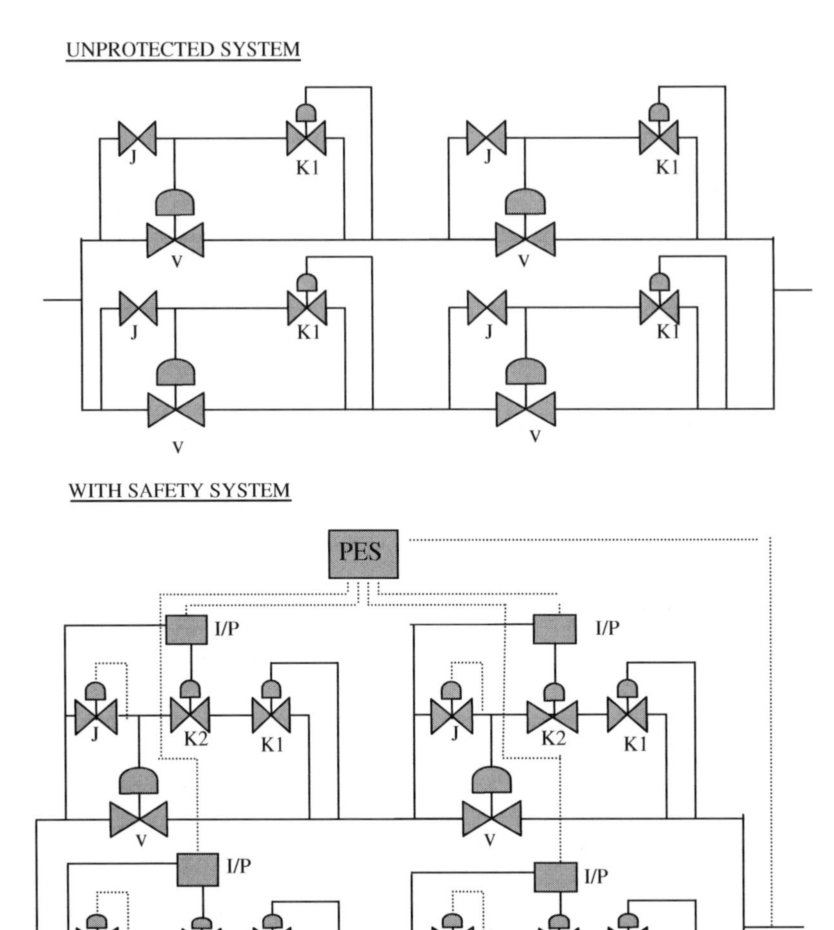

Figure 13.1: The system, with and without backup protection.

(b) It is credible that the existing pilots K1 and the add-on pilots K2 might have common cause failures. In that case some failures of K1 pilots would cause failure of their associated K2 pilots.

Therefore, in Chapter 11, a design is offered which does provide EUC/SRS independence. What then of the SIL target for the SRS in Figure 13.1?

It becomes necessary to regard the whole of the system as a single safety-related system. It thus becomes a high demand system with a Maximum Tolerable Failure Rate (see Chapter 11) of 10^{-5} pa. This is at the far limit of SIL 4 and is, of course, quite unacceptable. Thus an alternative design would be called for.

13.2 A Hand-held Alarm Intercom, Involving Human Error in the Mitigation

A rescue worker, accompanied by a colleague, is operating in a hazardous environment. The safety-related system, in this example, consists of a hand-held intercom intended to send an alarm to a supervisor should the user become incapacitated. In this scenario, the failure of the equipment (and lack of assistance from the colleague) results in the "alarm" condition not being received or actioned by a "supervisor" located adjacent to the hazardous area. This, in turn leads to fatality.

The scenario is modeled in Figure 13.2. Gate G1 models the demand placed on the safety related system and Gate G2 models the mitigation. The events:

ATRISK are the periods to which an individual is exposed
SEP is the probability that the colleague is unavailable to assist
HE1 is the probability that the colleague fails to observe the problem
INCAP is the probability that the colleague is incapacitated
DEMAND is the probability that the incident arises during the event
FATAL is the probability that the incident would lead to fatality if the worker is not rescued.

Assume that the frequency of Gate G1 is shown to be 4.3×10^{-4} pa. Assume, also, that the target Maximum Tolerable Risk is 10^{-5} pa. In order for the frequency of the top event to equal 10^{-5} pa the probability of failure associated with Gate G2 must be $1 \times 10^{-5} / 4.3 \times 10^{-4} = 2.33 \times 10^{-2}$. However the event HE2 has been assigned a PFD of 10^{-2}, which leaves the target PFD of the intercom to be $\mathbf{1.33 \times 10^{-2}}$.

Thus a **SIL 1 target (low demand)** will be placed on this safety function. Notice how critical the estimate of human error is in affecting the SIL target for the intercom. Had HE2 been 2×10^{-2} then the target PFD would have been $2.33 \times 10^{-2} - 2 \times 10^{-2} = 3.3 \times 10^{-3}$. In that case the target for the intercom would have **been SIL 2.**

13.3 Maximum Tolerable Failure Rate Involving Alternative Propagations to Fatality

In this example, as a result of instrument and plant failures, a toxic gas cloud is released. Two types of hazard are associated with the scenario:

(a) Concentration of Gas on Site

In this case a wind velocity of less than 1 m/sec is assumed, as a result of which inversion would cause a concentration of gas within the site boundary, possibly leading to fatality.

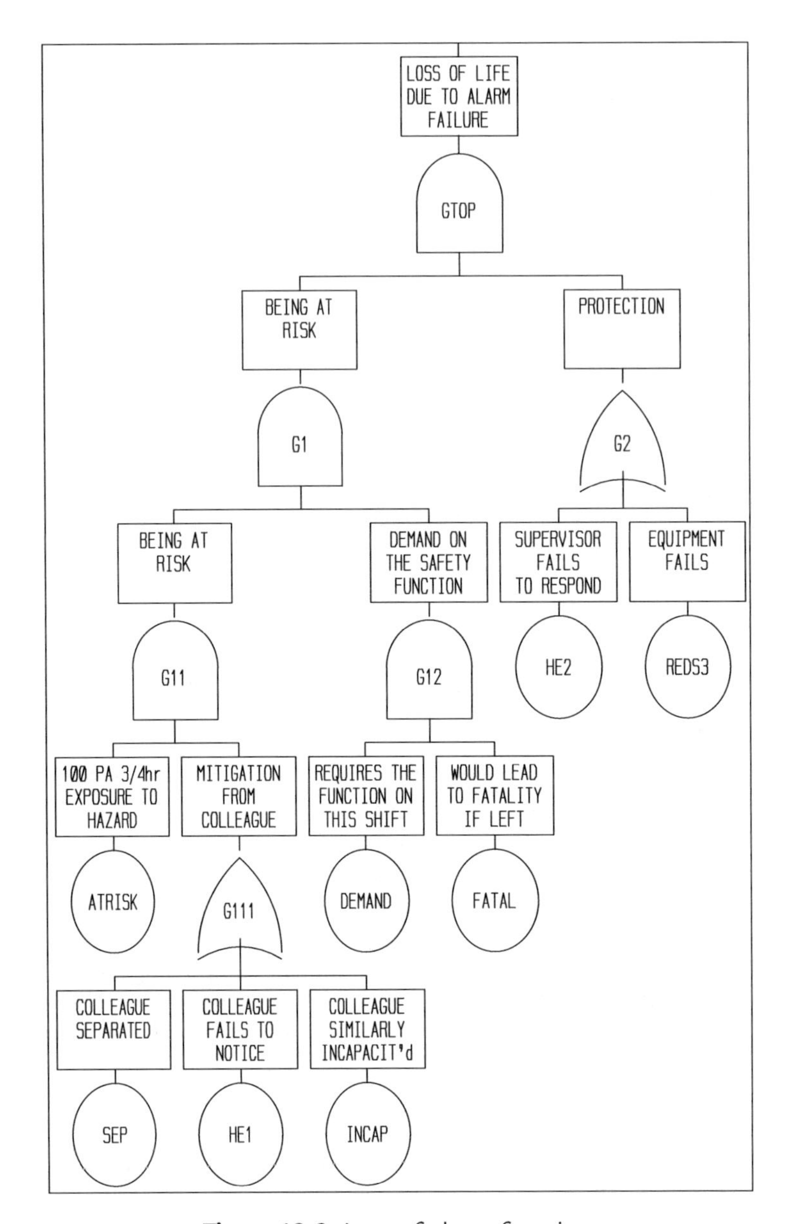

Figure 13.2: Loss of alarm function.

Max Tolerable Risk $= 10^{-5}$ pa (perhaps 10^{-4} pa overall voluntary risk but 10 similar hazards)
Downstream pipe rupture due to 8 Bar $= 10^{-2}$ pa
Wind < 1 m/sec assumed to be 1 day in 30 $= 3.3 \times 10^{-2}$
Plant in operation, thus causing exposure to the hazard, 100% of the time.
Personnel close enough = 75%

Propagation of failure to fatality is estimated to be 80%
Thus **Max Tolerable PFD** = 10^{-5} pa / (0.01 pa × 3.3 × 10^{-2} × 0.75 × 8)
= 5.1 × 10^{-2}

(b) Spread of Gas to Nearby Habitation

In this case a wind velocity of greater than 1 m/sec is assumed and a direction between north and north west, as a result of which the gas cloud will be directed at a significant area of population.

Max Tolerable Risk = 10^{-5} pa (public, involuntary risk)
Downstream pipe rupture due to 8 Bar = 10^{-2} pa
Wind > 1 m/sec assumed to be 29 days in 30 = 97%
Wind direction from E to SE, 15%
Plant in operation, thus causing exposure to the hazard, 100% of the time
Public present = 100%
Propagation of failure to fatality is assumed to be 20%
Thus **Max Tolerable PFD** = 10^{-5} pa / (0.01 pa × 0.97 × 0.15 × 0.20)
= 3.4 × 10^{-2}

The lower of the two **Max Tolerable PFDs is 3.4 × 10^{-2}**, which becomes the target.

SIL targets for the safety-related systems would be based on this. Thus, if only one level of protection were provided a **SIL 1 target** would apply.

13.4 Hot/cold Water Mixer Integrity

In this example, a programmable equipment mixes 70°C water with cold water to provide an appropriate outlet to a bath. In this scenario, a disabled person is taking a bath, assisted by a carer. The equipment failure, which leads to the provision of 70°C water, is mitigated by human intervention.

Figure 13.3 models the events leading to fatality. Gate G11 apportions the incidents between those failures occurring prior to the bath (such that it is drawn with scalding water)(G111) and those that occur during the bath (G112). It was assumed that a bath occupies § hr per 2 days. Thus the probability of the former is 47§ / 48 = 99% and the latter therefore 1%.

A 20% chance of a distraction arising is assumed.
A 10% chance of the carer responding to the distraction is assumed.
The human error whereby the carer fails to detect a scalding bath is estimated as **0.1.**

The reader might care to study Figure 13.3 and verify that the probability associated with gate G11 is (0.99 × [0.1 × 0.2 + 0.1]) + (0.01 × [0.1 × 0.2]) = **0.119.**

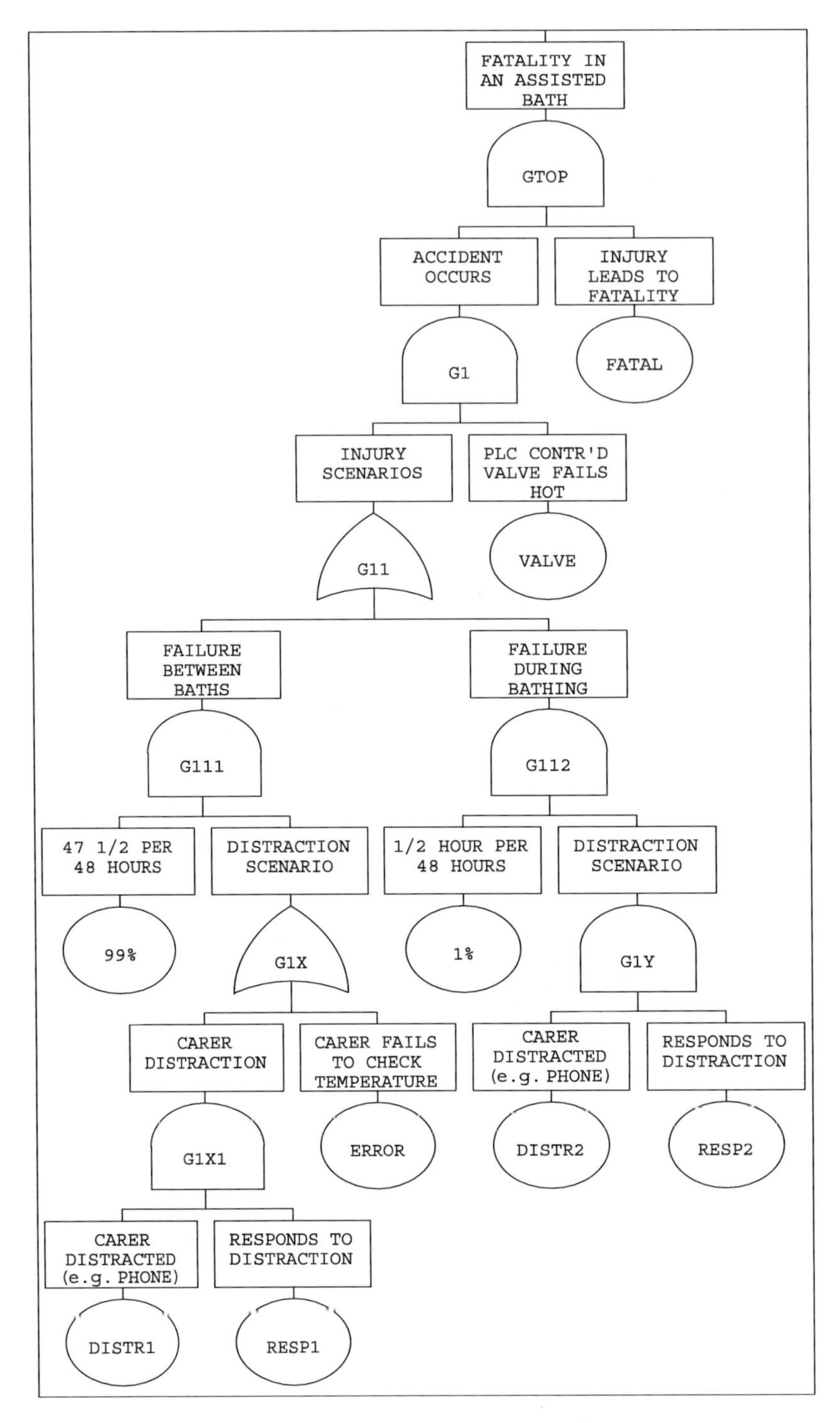

Figure 13.3: Fault tree – with assistance from a carer.

The probability of an incident becoming fatal has been estimated, elsewhere, as 8.1%. The maximum tolerable risk has been set as 10^{-5} pa, thus the maximum tolerable incident rate is 10^{-5} / 8.1% = **1.2×10^{-4} pa (Gate G1).**

The maximum tolerable failure rate for the product is therefore:

Gate G1/ Gate G11
= 1.2×10^{-4} pa / 0.119
= 1.01×10^{-3} pa.

This would imply a safety-integrity target of **SIL 2 (high demand).**

13.5 Scenario Involving High Temperature Gas to a Vessel

In this example, gas is cooled before passing from a process to a vessel. The scenario involves loss of cooling, which causes high temperature in the vessel, resulting in subsequent rupture and ignition. This might well be a three-fatality scenario.

Supply profile permits the scenario (pilot alight)	100%
Probability that drum ruptures	5%
Probability of persons in vicinity of site (pessimistically)	50%
Probability of ignition	90%
Probability of fatality	100%

Assuming a maximum tolerable risk of 10^{-5} pa, the maximum tolerable failure rate is 10^{-5} pa/ $(0.05 \times 0.5 \times 0.9)$ = **4.4×10^{-4} pa.**

The scenario is modeled in Figure 13.4. Only Gate G22 (involving human intervention and a totally independent equipment) is independent of the ESD (emergency shutdown system). If a probability of failure on demand in the SIL1 range (say 3×10^{-2}) is assigned to Gate G22 then the top event target reduces to 4.4×10^{-4} pa / 3×10^{-2} pa = 1.5×10^{-2} pa, which is also SIL1. Thus a **SIL 1 target (low demand)** is adequate for the ESD.

Assume that the frequency of the top event is **1.3×10^{-5} pa, which meets the target.**

ALARP

If a cost per life saved criteria of £4,000,000 is used then the expenditure on any proposal which might reduce the risk to 10^{-7} pa (based on 10^{-6} pa but with 10 similar hazards) can be calculated (based on a 30 year plant life) as:

The frequency of the top event maps to a risk of $1 \times 10^{-5} \times (1.3 \times 10^{-5} / 4.4 \times 10^{-4})$ = 3×10^{-7} pa and is thus in the ALARP region.
£ 4,000,000 = £proposed / ($[3 \times 10^{-7} - 1 \times 10^{-7}] \times 3$ deaths $\times$ 30 yrs)

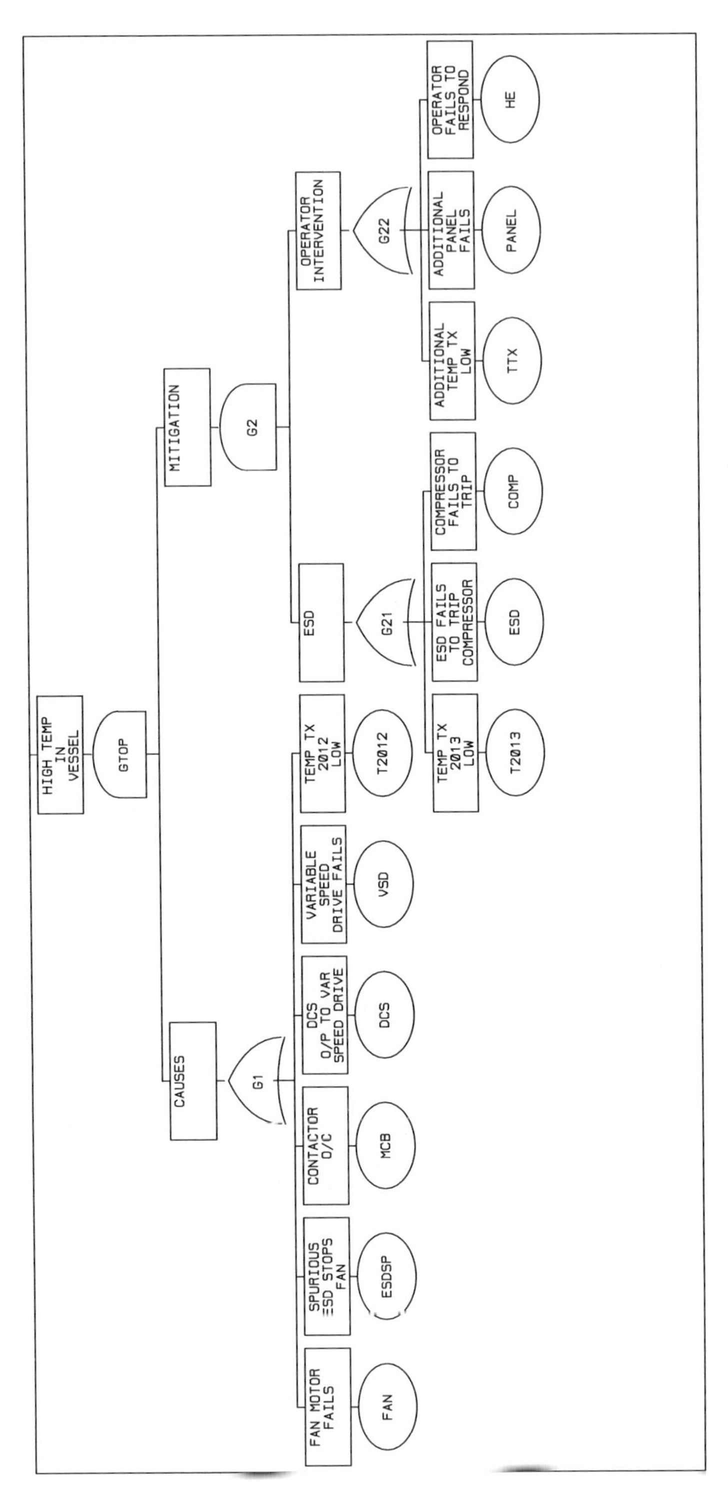

Figure 13.4: Fault tree — high temperature in vessel.

Thus £ proposed = £72
Any proposal involving less than £72, which would reduce the risk to 10^{-7} pa, should be considered. It is unlikely that any significant risk reduction can be achieved for that capital sum.

13.6 Example using the LOPA Technique

In Chapter 2.1.2 the LOPA (Levels of Protection Analysis) method was described. In this example, a Safety Integrity Level (SIL) assessment is conducted for a hydro-electric dam plant for the requirements of a Flood Gate Control System (FGCS). The required SIL is assessed for the control of the flood gates. These flood gates are required to prevent the dam from being overtopped when there is more water draining into the loch than the hydro turbines can use.

The major hazards identified are:

"Dam over-topping and a flood of water over ground that is used by ramblers"

and

"Water surge down the river which could cause a hazard to fishermen standing in the river".

Assignment of SIL requirements: the objective is to review the specified hazards and provide a quantitative assessment of the levels of risk reduction required in addition to the existing controls.

Current controls: there is remote control from a central control room, via communication links, to an independent SIL 2 remote manual flood gate control system.

There is also an independent local control panel which will provide a local manual facility to open/close the gate.

The LOPA analysis is to determine the functional safety requirements for a local automatic flood control system.

SIL targeting: Table 13.1 summarizes the LOPA and the required Probability of Failure on Demand (PFD) values and corresponding SILs for each hazard.

The assessment of whether the targets are met is carried out in Chapter 16.1.

The LOPA Worksheets are presented below. Notice how the PFD, which determines the target SIL, is obtained, in each worksheet, from ratio of the "Maximum tolerable risk" to the column called "Intermediate Event Likelihood" (actually a frequency).

Table 13.1: Summary of the LOPA.

Event (hazard) description	Consequence	Safety Instrumented Function (SIF) requirement (PFD)	SIF requirement (SIL)	SIF description
Dam over-topping due to gates failing to open on demand during a major storm (requiring the use of 1 gate), which spillways are unable to mitigate	Death of more than one person	5.0×10^{-3}	SIL2	PLC to provide independent automatic control of flood gates to open gates when there are flood conditions
Water surge: gates open spuriously causing a surge of water which could drown multiple fishermen	Death of more than one person	2.3×10^{-3}	SIL2	Watchdog to monitor the gate drive outputs from the PLC and if required disable outputs

Table 13.2: LOPA Worksheet — Dam over-topping.

Event (hazard) description	Consequence	Maximum tolerable risk (/yr)	Initiating cause	Initiating likelihood (/yr)	IPLs: Vulnerability: e.g.: probability of affectation, direction of release, wind	IPLs: General purpose design: e.g.: additional mechanical safety margin	IPLs: Basic control system (BCS): e.g. independent control system, alarms	IPLs: Additional control systems (independent of BCS)	IPLs: Alarms (independent of BCS)	IPLs: Additional mitigation - access: e.g. usage, restricted access, occupancy, fences, avoidance	IPLs: Additional mitigation - procedural: e.g. operator action, detection, inspections	IPLs: Additional mitigation - physical: e.g. alternative physical protection, spill ways etc.	Intermediate event likelihood	SIF requirement (PFD)	SIF requirement (SIL)
				[a]	[b]	[c]	[d]	[e]	[f]	[g]	[h]	[j]			
Dam over-topping due to gates failing to open on demand during a major storm (requiring the use of 1 gate), which spillways are unable to mitigate.	Death of more than one person	1.00E-06	Adverse weather	1	1	1	1	1	0.01	0.2	0.1	1	2.00E-04	5.00E-03	SIL2
				Storms severe enough to require the use of 1 gate occur once per year					Various weather/river level warnings available to operator in Central Control Room – other parts of river will be rising, providing extra warning. Credit based on analysis of communications, and operator training / experience	From surveys it is estimated that there is less than 20% probability that the general public will be in the area during the adverse weather conditions	Local operator presence during storms – gates can be opened using mechanical winder or power assisted drive. If a mechanical failure of the gate has occurred, the operator could open a different gate				

Table 13.3: LOPA Worksheet — Water Surge.

Event (hazard) description	Consequence	Maximum tolerable risk (/yr)	Initiating cause	Initiating likelihood (/yr)	Vulnerability: e.g.: probability of affectation, direction of release, wind	IPLs: General purpose design: e.g.: additional mechanical safety margin	IPLs: Basic control system (BCS): e.g. independent control system, alarms	IPLs: Additional control systems (independent of BCS)	IPLs: Alarms (independent of BCS)	IPLs: Additional mitigation - access: e.g. usage, restricted access, occupancy, fences, avoidance	IPLs: Additional mitigation - procedural: e.g. operator action, detection, inspections.	IPLs: Additional mitigation- physical: e.g. alternative physical protection, spill ways etc.	Intermediate event likelihood	SIF requirement (PFD)	SIF requirement (SIL)
				[a]	[b]	[c]	[d]	[e]	[f]	[g]	[h]	[j]			
Water surge: Gate opening spuriously causing a surge of water which could drown multiple fishermen	Death of more than one person	1.00E-06	Output relay / contactor circuit fails closed	1.00E-02	1	1	0.01	1	1	0.21	1	1	2.1 E-05	2.3 E-03	SIL2
				Failure rate of armature relay (30% dangerous – contact S/C)			SIL 2 assessed			Fishing season lasts for 8 months per year. Fishing 15 hours per day. Estimated fishing takes place 50% of possible time					
			Flood gate control PLC fails to danger, causing a gate to open at double-speed.	2.00 E-03	1	1	1	1	1	0.21	1	1	4.2 E-04		
				Rate at which either FG PLC energize the output contactor or open relay spuriously (and thus causes a gate to open at double-speed)						Fishing season lasts for 8 months per year. Fishing 15 hours per day. Estimated fishing takes place 50% of possible time					
			Total										**4.4 E-04**		

CHAPTER 14

Hypothetical Rail Train Braking System (Example)

Chapter Outline

The following example has been simplified and, as a consequence, some of the operating modes have been changed in order to maintain the overall philosophy but give clarity to the example.

14.1 The Systems

In this example we have a combination of two safety-related systems. One is a "high demand" train primary braking system, together with a second level of protection consisting of a "low demand" emergency braking system.

Typically there are at least two methods of controlling the brakes on carriage wheels. The "high demand" system would be the primary braking function activated by either the train driver or any automatic signaled input (such as ATP). This system would send electronic signals to operate the brakes on each bogie via an air-operated valve. This is a proportional signal to regulate the degree of braking. The system is normally energized to hold brakes off. The output solenoid is de-energized to apply the brakes.

Each bogie has its own air supply reservoir topped up by an air generator. Air pressure has to be applied to operate the brakes. However, each bogie braking system is independent and each train has a minimum of two carriages. The loss of one bogie braking system would reduce braking by a maximum of 25%. It is assumed that the safety function is satisfied by three out of the four bogies operating (i.e. two must fail).

Safety Critical Systems Handbook. DOI: 10.1016/B978-0-08-096781-3.10014-8

In addition to this primary braking system there is separate emergency braking. This is a single electrical wire loop that runs the full length of the train connected to an emergency button in the driver's cab. This circuit operates a normally energized solenoid valve. This circuit holds the brakes off and the emergency solenoids are de-energized to apply full braking pressure to the brakes.

Figure 14.1 shows the general arrangement of the two systems serving four bogies over two carriages.

14.2 The SIL Targets

The specification for this design requires a **SIL 2 target for the primary braking** system, and a **SIL 3 target for the emergency braking** system.

These targets may have been arrived at by a risk graph approach. Therefore, unlike Chapter 11 where a specific quantified target was assessed, the SIL targets only provide an order of magnitude range of failure rates (or probabilities of failure on demand) for each of the two safety-related systems.

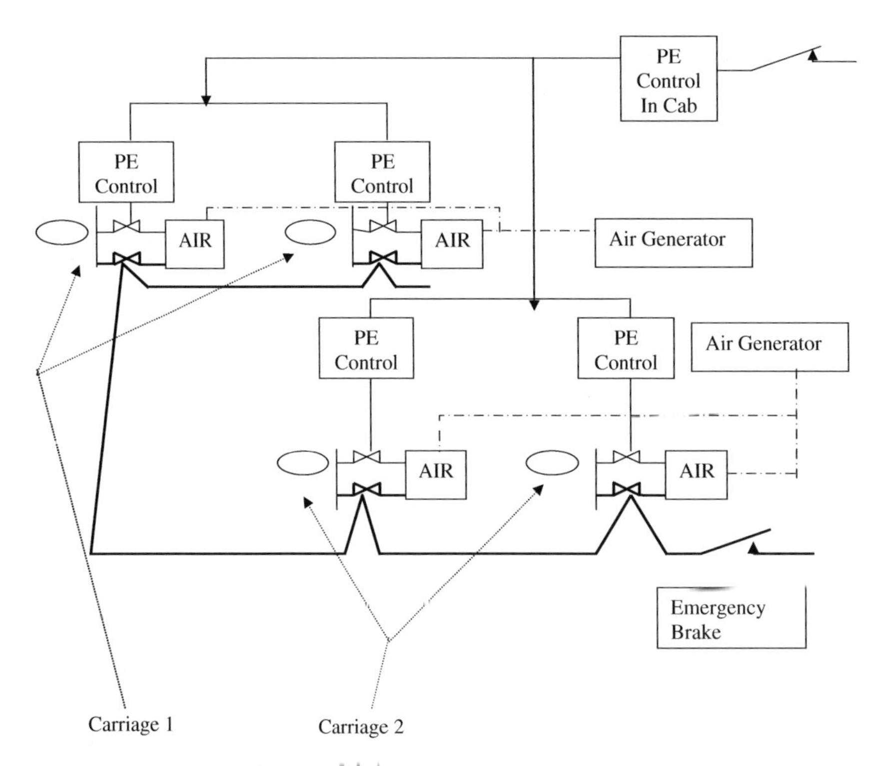

Figure 14.1: Braking arrangement.

The SIL 2 braking system is a high demand system and, thus, the target is that the failure rate is **less than 10^{-2} pa.**

The SIL 3 emergency braking system is a low demand system and, thus, the target is that the probability of failure on demand is **less than 10^{-3}.**

It should be noted that the two systems are not independent in that they share the air power and brake actuator systems. As a result the overall safety-integrity cannot be assessed as the combination of independent SIL 2 and SIL 3 systems. The common elements necessitate that the overall integrity is assessed as a combination of the two systems and this will be addressed in section 14.6.

14.3 Assumptions

As in Chapter 11, assumptions are key to the validity of any reliability model and its quantification.

(a) Failure rates (symbol λ), for the purpose of this prediction, are assumed to be constant with time. Both early and wearout-related failures are assumed to be removed by burn-in and preventive replacement respectively.
(b) The majority of failures are revealed on the basis of 2 hourly usage. Thus, half the usage interval (1 hour) is used as the down time.
(c) The proof-test interval of the emergency brake lever is 1 day. Thus the average down time of a failure will be 12 hours.
(d) The common cause failure beta factor will be determined by the same method as in Chapter 11. A partial beta factor of 1% is assumed, for this example, in view of the very high inspection rate.
(e) The main braking cab PE controller operates via a digital output. The bogie PE operates the valve via an analogue output.

14.4 Failure Rate Data

Credible failure rate data for this example might be:

Item	Failure mode	Failure rates (10^{-6} per hour)		MDT (hrs)
		Total	Mode	
PES (cab)	Serial output low	2	0.6	1
PES (bogie)	Analogue ouput low	2	0.6	1
Actuated valve	Fail to move	5	1.5	1
Solenoid valve	Fail to open	0.8	0.16	
Driver's levers				

(Continued)

Item	Failure mode	Failure rates (10^{-6} per hour)		MDT (hrs)
		Total	Mode	
Emergency	Fail to open contact	1	0.1	12
Main	No braking	1	0.1	1
Bogie air reservoir system (reservoir check valve and compressor) achieved by regular (daily use)	Fail	1	1	1
Brake shoes A low failure rate achieved by regular (2 weeks) inspection	Fail	0.5	0.5	1
Common cause failure of Air			0.05	
Common cause failure of brake shoes			0.005	

14.5 Reliability Models

It is necessary to model the "top event" failure for each of the two systems. Chapter 11 used the reliability block diagram method and, by contrast, this chapter will illustrate the fault tree approach.

14.5.1 Primary Braking System (High Demand)

Figure 14.2 is the fault tree for failure of the primary braking system. Gates G22 and G23 have been suppressed to simplify the graphics. They are identical, in function, to G21 and G24. Note that the Gate G2 shows a figure "2", being the number of events needed to fail.

The frequency of the top event is **6.6×10^{-3} pa, which meets the SIL 2 target.**

The table below the fault tree in Figure 14.2 shows part of the fault tree output from the Technis TTREE package (see end of book). The cutsets have been ranked in order of frequency since this is a high demand scenario which deals with a failure rate. Note that 80% of the contribution to the top event is from the PE1 event.

14.5.2 Emergency Braking System (Low Demand)

Figure 14.3 is the fault tree for failure of the emergency braking system. Gates G22 and G23 have been suppressed in the same way as for Figure 14.2.

The probability of the top event is **1.3×10^{-6}, which meets the SIL 3 target with approximately 2 orders of magnitude margin.**

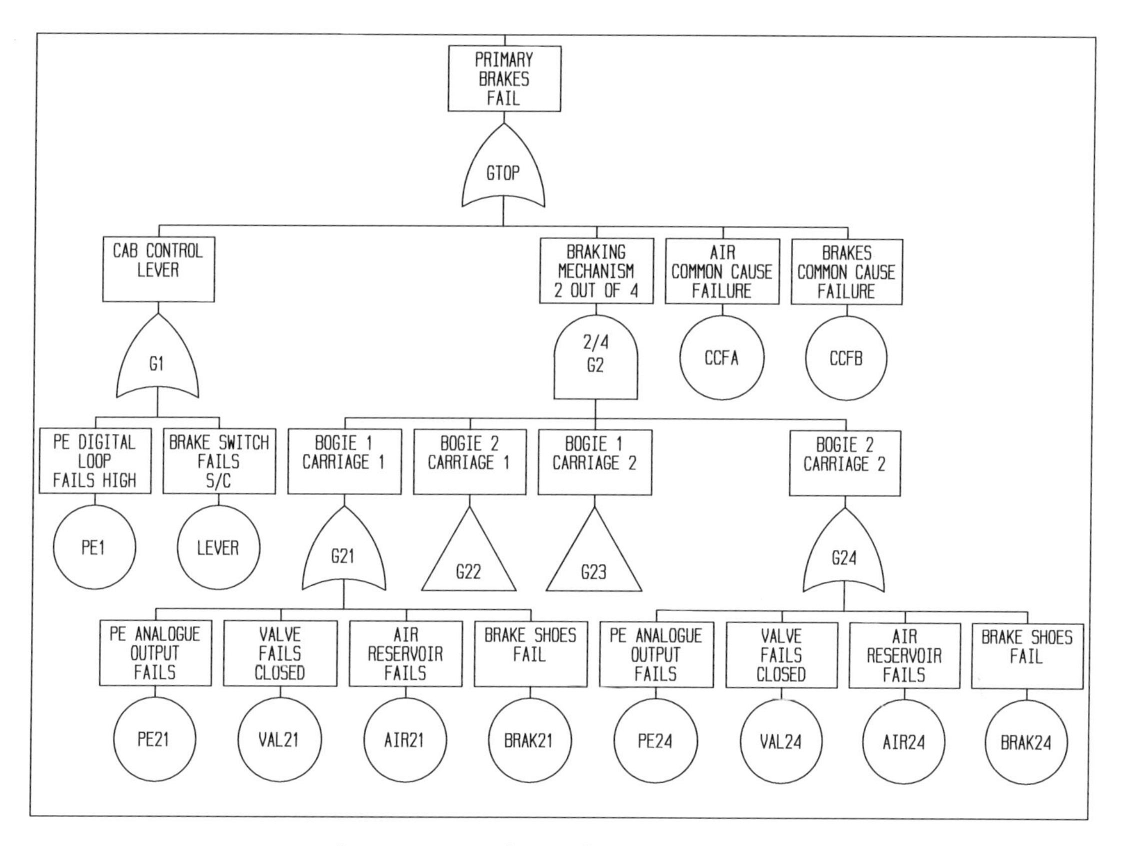

Figure 14.2: Fault tree for primary braking.

The table below the fault tree in Figure 14.3 shows part of the fault tree output as in the previous section. In this case the cutsets have been ranked in order of probability since this is a low demand scenario which deals with a PFD. Note that >95% of the contribution to the top event is from the EMERG event (lever).

14.6 Overall Safety Integrity

As mentioned in section 14.2 the two safety-related systems are not independent. Therefore the overall failure rate (made up of the failure rate of the primary braking and the PFD of the emergency braking) is calculated as follows. The fault tree in Figure 14.4 combines the systems and thus takes account of the common elements in its quantification.

The overall failure rate is $\mathbf{4.8 \times 10^{-4}}$ **pa.** The cutset rankings show that the air supply Common Cause Failure accounts for 90% of the failures.

```
Results of fault tree quantification for top event: GTOP
Top event frequency =    0.755E-06 per hour
                         0.662E-02 per year
Top event MTBF      =    0.132E+07 hours
                         0.151E+03 years
Top event MDT       =    0.100E+01 hours
Top event probability =  0.755E-06
--------------------------------------------
Basic Event Reliability Data

Basic     Type    Failure    Mean Fault    Constant
Event              Rate       Duration    Probability
CCFA       I/E    .500E-07      1.00
CCFB       I/E    .500E-08      1.00
PE1        I/E    .600E-06      1.00
LEVER      I/E    .100E-06      1.00
PE21       I/E    .600E-06      1.00
VAL21      I/E    .150E-05      1.00
AIR21      I/E    .100E-05      1.00
BRAK21     I/E    .500E-06      1.00
PE22       I/E    .600E-06      1.00
VAL22      I/E    .150E-05      1.00
AIR22      I/E    .100E-05      1.00
BRAK22     I/E    .500E-06      1.00
PE23       I/E    .600E-06      1.00
VAL23      I/E    .300E-05      1.00
AIR23      I/E    .100E-05      1.00
BRAK23     I/E    .500E-06      1.00
PE24       I/E    .600E-06      1.00
VAL24      I/E    .150E-05      1.00
AIR24      I/E    .100E-05      1.00
BRAK24     I/E    .500E-06      1.00
--------------------------------------------
Barlow-Proschan measure of cut set importance

Rank    1   Importance .795       MTBF hours.167E+07   MTBF years 190.

        Basic     Type    Failure    Mean Fault    Constant
        Event              Rate       Duration    Probability
        PE1        I/E    .600E-06     1.00

Rank    2   Importance .132       MTBF hours.100E+08   MTBF years .114E+04

        Basic     Type    Failure    Mean Fault    Constant
        Event              Rate       Duration    Probability
        LEVER      I/E    .100E-06     1.00

Rank    3   Importance .662E-01   MTBF hours.200E+08   MTBF years .228E+04

        Basic     Type    Failure    Mean Fault    Constant
        Event              Rate       Duration    Probability
        CCFA       I/E    .500E-07     1.00

Rank    4   Importance .662E-02   MTBF hours.200E+09   MTBF years .228E+05

        Basic     Type    Failure    Mean Fault    Constant
        Event              Rate       Duration    Probability
        CCFB       I/E    .500E-08     1.00
```

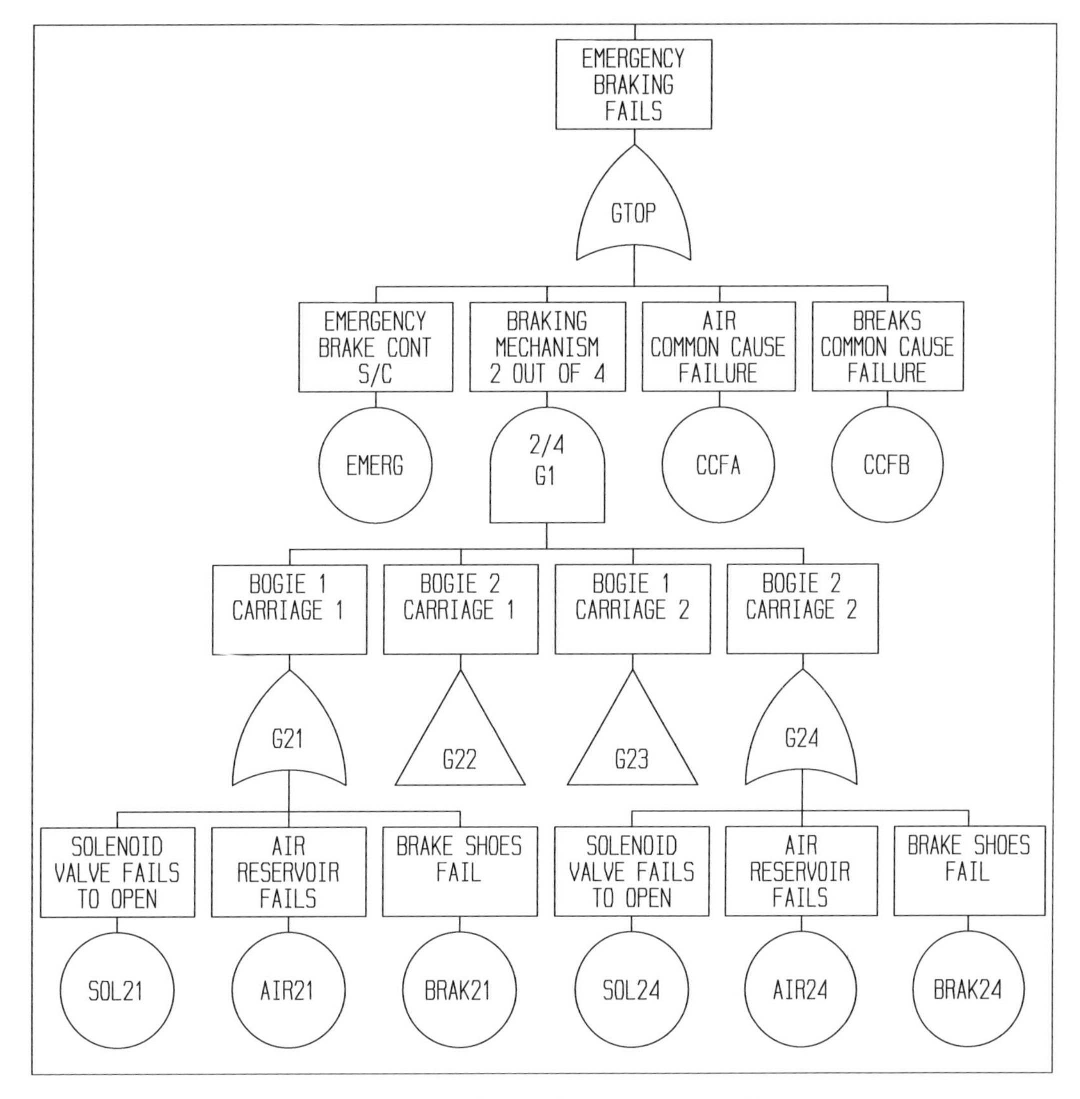

Figure 14.3: Fault tree for emergency braking.

This example emphasizes that, since the two systems are not independent, one cannot multiply the failure rate of the primary braking system (6.6×10^{-3} pa) by the PFD of the emergency braking system (3.6×10^{-6}). The result would be nearly 4 orders optimistic and the overall arrangement has to be modeled as shown in Figure 14.4

```
Results of fault tree quantification for top event: GTOP
Top event frequency =     0.155E-06 per hour
                          0.136E-02 per year
Top event MTBF      =     0.645E+07 hours
                          0.736E+03 years
Top event MDT       =     0.809E+01 hours
Top event probability =   0.126E-05
--------------------------------------------
Basic Event Reliability Data
Basic     Type    Failure    Mean Fault    Constant
Event              Rate       Duration    Probability
EMERG      I/E    .100E-06      12.0
CCFA       I/E    .500E-07      1.00
CCFB       I/E    .500E-08      1.00
SOL21      I/E    .160E-06      12.0
AIR21      I/E    .100E-05      1.00
BRAK21     I/E    .500E-06      1.00
SOL22      I/E    .160E-06      12.0
AIR22      I/E    .100E-06      1.00
BRAK22     I/E    .500E-05      1.00
SOL23      I/E    .160E-06      12.0
AIR23      I/E    .100E-05      1.00
BRAK23     I/E    .500E-06      1.00
SOL24      I/E    .160E-06      12.0
AIR24      I/E    .100E-05      1.00
BRAK24     I/E    .500E-06      1.00
--------------------------------------------
Fussell-Vesely measure of cut set importance

Rank    1   Importance .956       Cut set probability .120E-05

        Basic     Type    Failure    Mean Fault    Constant
        Event              Rate       Duration    Probability
        EMERG      I/E    .100E-06    12.0

Rank    2   Importance .398E-01   Cut set probability .500E-07

        Basic     Type    Failure    Mean Fault    Constant
        Event              Rate       Duration    Probability
        CCFA       I/E    .500E-07    1.00

Rank    3   Importance .398E-02   Cut set probability .500E-08

        Basic     Type    Failure    Mean Fault    Constant
        Event              Rate       Duration    Probability
        CCFB       I/E    .500E-08    1.00

Rank    4   Importance .765E-05   Cut set probability .960E-11

        Basic     Type    Failure    Mean Fault    Constant
        Event              Rate       Duration    Probability
        SOL21      I/E    .160E-06    12.0
        BRAK22     I/E    .500E-05    1.00
```

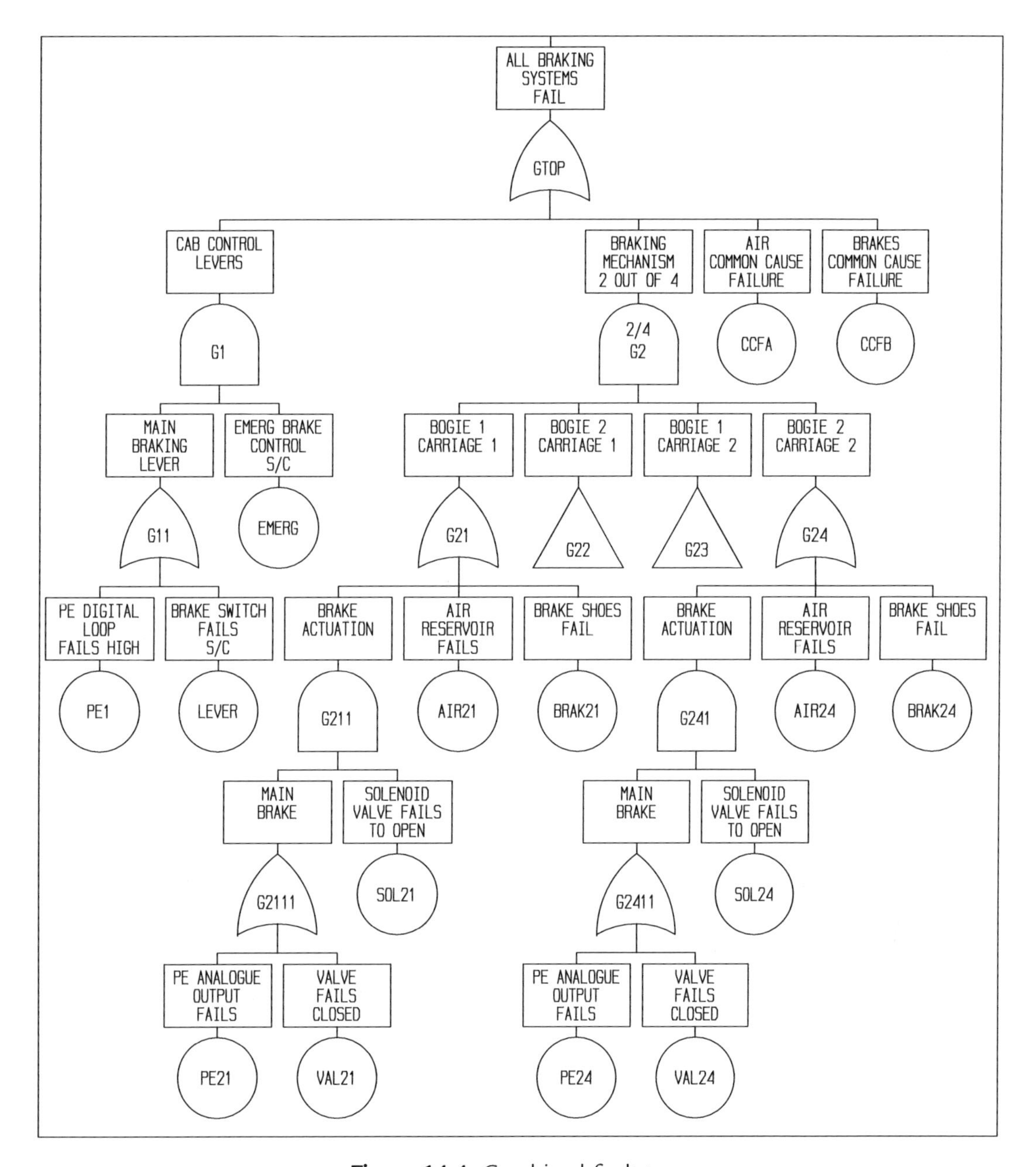

Figure 14.4: Combined fault tree.

```
Results of fault tree quantification for top event: GTOP
Top event frequency =    0.550E-07 per hour
                         0.482E-03 per year
Top event MTBF      =    0.182E+08 hours
                         0.207E+04 years
Top event MDT       =    0.100E+01 hours
Top event probability =  0.550E-07
-------------------------------------------
Basic Event Reliability Data
Basic     Type    Failure    Mean Fault    Constant
Event              Rate       Duration    Probability
CCFA       I/E    .500E-07       1.00
CCFB       I/E    .500E-08       1.00
EMERG      I/E    .100E-06       12.0
PE1        I/E    .600E-06       1.00
LEVER      I/E    .100E-06       1.00
AIR21      I/E    .100E-05       1.00
BRAK21     I/E    .500E-06       1.00
SOL21      I/E    .160E-06       12.0
PE21       I/E    .600E-06       1.00
VAL21      I/E    .150E-05       1.00
AIR22      I/E    .100E-05       1.00
BRAK22     I/E    .500E-06       1.00
SOL22      I/E    .160E-06       12.0
PE22       I/E    .600E-06       1.00
VAL22      I/E    .150E-05       1.00
AIR23      I/E    .100E-05       1.00
BRAK23     I/E    .500E-06       1.00
SOL23      I/E    .160E-06       12.0
PE23       I/E    .600E-06       1.00
VAL23      I/E    .300E-05       1.00
AIR24      I/E    .100E-05       1.00
BRAK24     I/E    .500E-06       1.00
SOL24      I/E    .160E-06       12.0
PE24       I/E    .600E-06       1.00
VAL24      I/E    .150E-05       1.00
-------------------------------------------
Barlow-Proschan measure of cut set importance

Rank    1   Importance .909       MTBF hours.200E+08   MTBF years .228E+04

        Basic    Type    Failure    Mean Fault    Constant
        Event             Rate       Duration    Probability
        CCFA      I/E    .500E-07    1.00

Rank    2   Importance .909E-01   MTBF hours.200E+09   MTBF years .228E+05

        Basic    Type    Failure    Mean Fault    Constant
        Event             Rate       Duration    Probability
        CCFB      I/E    .500E-08    1.00

Rank    3   Importance .363E-04   MTBF hours.500E+12   MTBF years .571E+08

        Basic    Type    Failure    Mean Fault    Constant
        Event             Rate       Duration    Probability
        AIR21     I/E    .100E-05    1.00
        AIR22     I/E    .100E-05    1.00
```

CHAPTER 15

Rotorcraft Accidents and Risk Assessment

Chapter Outline

This chapter is in two parts. The first presents some helicopter accident and fatality statistics to provide a comparison with other activities. The second is an example of a quantified risk assessment, which required an input from the foregoing statistics.

15.1 Helicopter Incidents

The following statistics are based on the Robinson R22 and R44 machines, which account for 25% of the rotorcraft flying hours in the UK (based on CAA statistics 1996–2009). They are used mainly for training and private helicopter flights.

The incident data obtained from the AAIB (Air Accident Investigation Bureau) cover just over 800,000 flying hours (1996–2008). The following table summarizes the rates for R22s and R24s.

	R22	R44
Incidents per craft operating hour	**1 in 6,000** equivalent to **30 m/c years**	**1 in 6,100** equivalent to **55 m/c years**
Injuries per craft operating hour	**1 in 19,000** (relatively **minor)**	**1 in 15,000 (**more **serious)**
Fatalities FAFR	**980 per 100 M hrs**	**1600 per 100 M hrs**
Injuries which are fatal	**15%**	**23%**

Table 15.1 compares the risk of fatality with a number of other activities. The term FAFR refers to the number of fatalities per 100,000 hours (10^8 hrs) of exposure to the activity in question. For voluntary activities (such as those listed) where exposure varies between activities it is a more representative metric than fatalities per calendar year (which take no account of relative

Safety Critical Systems Handbook. DOI: 10.1016/B978-0-08-096781-3.10015-X

Table 15.1: Fatal accident frequency (FAFR).

Activity	FAFR (per 100 million hours exposure)
Staying at home	2
Train/bus	4
Motor car	50
Coal mining	21
Trawler (fishing)	60
Civil air travel	120
Helicopter North Sea offshore UK	***200***
Canoe	400
Helicopter offshore worldwide	***640***
Motorcycle	800
R22 (1996–2009) UK	**980**
Swimming	1,300
All helicopters (USA)	***1,480***
R44 (1997–2009) UK	**1,600**
Gliding	2,900
Rock climbing	4,000
Boxing	20,000
Horse riding	28,000

exposure). The comparative FAFRs were obtained from Reliability, Maintainability and Risk, 7th Edition, D J Smith, Elsevier (Butterworth Heinemann), ISBN 07506 6694 3, and from various web searches.

Of the fatal helicopter accidents (10 in total) three were due to deliberate flight in weather conditions which private pilots are taught to avoid. These fatalities can be argued to be totally avoidable and, thus, recalculating the FAFRs leads to the shortened Table 15.2.

Table 15.2: Revised FAFRs.

Activity	FAFR (per 100 million hours exposure)
Civil air travel	120
Helicopter North Sea offshore UK	***200***

(Continued)

Table 15.2 *continued*

Activity	FAFR (per 100 million hours exposure)
R44 (1996–2007) UK	**600**
Helicopter offshore worldwide	***640***
R22 (1997–2007) UK	**730**
Motorcycle	800
All helicopters (USA)	***1,480***
Gliding	2,900
Rock climbing	4,000

The picture which emerges is of an activity which is far safer than popular perception.

15.2 Floatation Equipment Risk Assessment

The following risk assessment involves "testing" the proposal to fit mandatory automatic floatation equipment to the skids of helicopters against the principles outlined in this book and to apply the ALARP principle to the findings.

It must be stressed that this study was based on assumptions made by the author and that therefore the findings might well alter if those assumptions were to be challenged.

In this scenario, a forced landing on water leads to fatality due to life jackets and/or floatation equipment being ineffective. A two-fatality scenario is assumed. The maximum individual tolerable risk target is taken 10^{-4} fatalities pa.

15.2.1 Assessment of the Scenario

The frequency of ditching event was obtained from the studies referred to in the previous section 15.1. Other frequencies, probabilities and costs were obtained by discussions within the aircraft industry.

The fault tree (Figure 15.1) was constructed to model the scenario without the benefit of floatation equipment. It was analysed using the TECHNIS fault tree package TTREE. The frequency of the top event is **7.2×10^{-5} pa**, which meets the target.

Figure 15.2 shows the modified fault tree, which credits the mitigation offered by floatation equipment. The frequency of the top event is **1.5×10^{-5} pa**, which also meets the target.

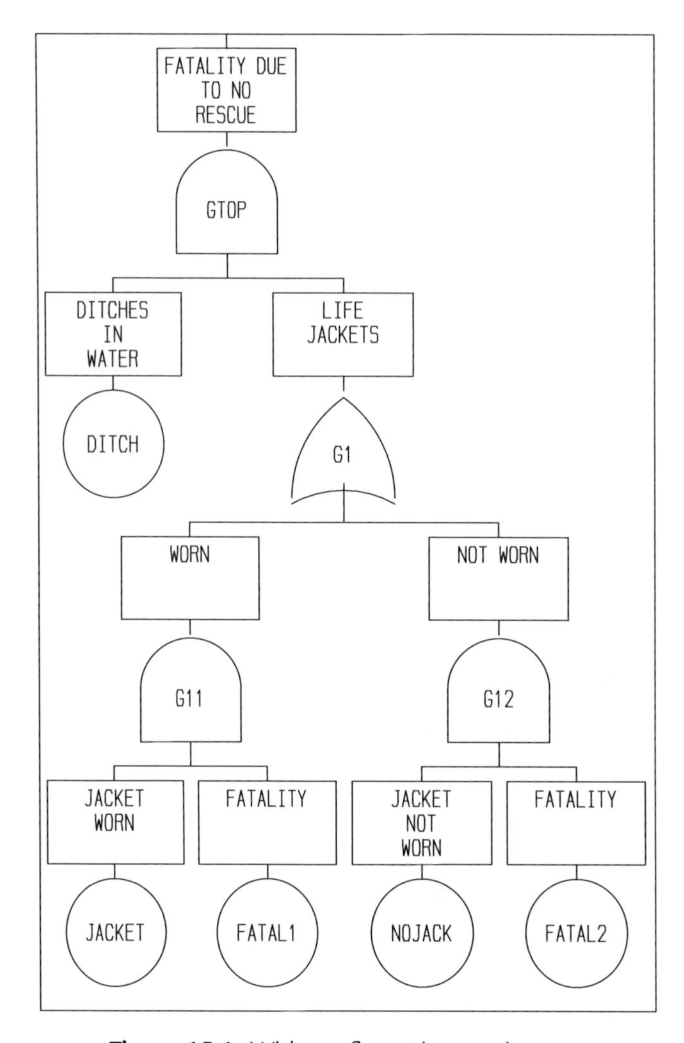

Figure 15.1: Without floatation equipment.

15.2.2 ALARP

Assuming the cost of a floatation system is £17,000 and assuming a 10-year equipment life then the cost per life saved arising from the risk reduction is:

$$\pounds 17,000/(7.2 \times 10^{-5} - 1.5 \times 10^{-5}) \times 2 \times 10 = \pounds 15 \text{ million pounds}$$

Since this exceeds the criterion mooted in chapter 2.2, ALARP could be argued to be satisfied without the additional risk reduction.

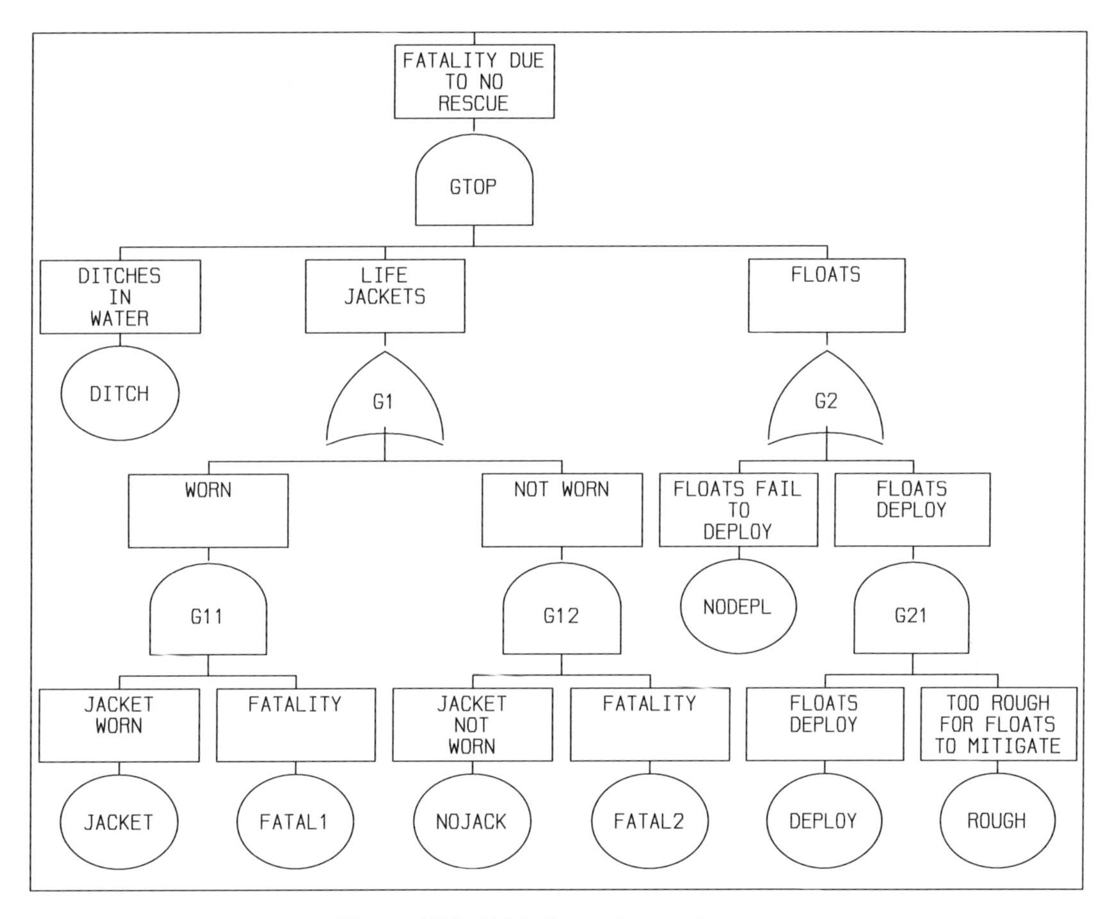

Figure 15.2: With floatation equipment.

CHAPTER 16

Hydro-electric Dam and Tidal Gates

Chapter Outline

16.1 Flood-gate Control System

16.1.1 Targets

This example provides a Safety Integrity Level (SIL) assessment of the proposed flood gate control system (FGCS) at a hydro-electric dam, demonstrating that it meets the identified hardware reliability and minimum configuration requirements in accordance with IEC 61508.

In order to identify the SIL requirements, a Layer of Protection Analysis (LOPA) was conducted at a meeting of interested parties. The study considered the hydro-electric plant to determine potential risks associated with the specified hazards. See example in Chapter 13.6.

Table 16.1 summarizes the LOPA and the required Probability of Failure on Demand (PFD) values and corresponding SILs for each of the two hazards.

The FGCS was then analysed to identify the SIFs used to mitigate the specified hazards, as presented in Table 16.2.

Safety Critical Systems Handbook. DOI: 10.1016/B978-0-08-096781-3.10016-1

Table 16.1: Summary of the LOPA.

Event (hazard) description	Consequence	Safety Instrumented Function (SIF) requirement (PFD)	SIF requirement (SIL)
Dam over-topping due to gates failing to open on demand during a major storm (requiring the use of 1 gate), which spillways are unable to mitigate	Death of more than one person	5.0×10^{-3}	SIL2
Water surge: gates open spuriously at full speed, causing a surge of water which could drown multiple fishermen	Death of more than one person	2.3×10^{-3}	SIL2

16.1.2 Assessment

(a) Common cause failures (CCFs)

The β values used in the analysis were based on engineering judgement and are presented in Table 16.3.

(b) Assumptions

The following summarizes the general assumptions used in the assessment:

the FGCS is assumed to be a low demand systems and therefore the LOW DEMAND PFD targets apply;
the analysis assumes that all failure modes that are not revealed by self test will be identified by the proof test, i.e. the proof test is 100% effective;

Table 16.2: Summary of safety functions.

Loop ref.	Input device	Input config.	Logic device	Logic config.	Output device	Output config.	Safety function
A	Level transmitters microwave (2 off) / radar	2oo3	Safety PLC	1oo1	Two flood gate drives	1oo2	Detection of high loch level opens 1 out of 2 (1oo2) floodgates
B	Safety timer relay	1oo1	N/A	N/A	Line contactor	1oo1	If the open contactor is closed for more than 50 seconds (i.e. the gate is opening too quickly), power is isolated from the motor by opening the line contactor

Table 16.3: CCF contributions

Redundant configuration	CCF β-factor	Justification
Microwave / radar level transmitters	5%	Three devices are mounted with separation and ultilize two dissimilar technologies
Flood gate operation mechanism	2%	The flood gates (and the associated lifting gear) are physically separated from one another
Power supplies	10%	The two supplies are of similar technology

the calculation of PFD is based upon an assumed MTTR of 24 hours;
if a failure occurs, it is assumed that on average it will occur at the mid point of the test interval; in other words, the fault will remain undetected for 50% of the test period;
the analysis assumes constant failure rates and therefore the effects of early failures are expected to be removed by appropriate processes; it is also assumed that items are not operated beyond their useful life, thus ensuring that failures due to wear-out mechanisms do not occur

(c) Failure rates of component parts

Table 16.4 summarizes the data sources.

Table 16.4: Failure rates and the calculation of SFF.

Item / function	Dangerous failure mode	λDD	λDU	λS	SFF	Source
DC motor	Fails to start on demand	0.0E+00	1.8E-06	3.3E-06	65%	Faradip v.6.1
Motor brake	Fails on	0.0E+00	8.4E-08	3.6E-08	30%	NRPD-85
Chain drive	Breaks	0.0E+00	2.7E-06	3.0E-07	10%	Faradip v.6.1
Redundant power supply	Loss of power	5.5E-05	0.0E+00	0.0E+00	100%	Faradip v.6.1
Microwave level transmitter	Fails to detect high loch level	9.9E-07	2.0E-07	3.4E-07	87%	Manufacturer's data adjusted, see Chapter 6.2.4
FG PLC AI module	Fails to interpret high loch level	5.6E-07	2.1E-07	4.2E-07	82%	ESC Failure Rate Database
Radar level transmitter	Fails to detect high loch level	1.1E-06	3.6E-07	4.7E-07	82%	Manufacturer's data adjusted, see Chapter 6.4
Resolver	Erroneously detects gate in open position	1.4E-06	1.5E-07	1.5E-06	95%	Faradip v.6.1

(*Continued*)

Table 16.4 ***continued***

Item / function	Dangerous failure mode	λDD	λDU	λS	SFF	Source
FG PLC AI module	Erroneously detects gate in open position	5.6E-07	2.1E-07	4.2E-07	82%	ESC Failure Rate Database
FG PLC CPU	Fails to interpret high level or gate closed on demand	2.7E-07	3.0E-08	2.6E-06	99%	ESC Failure Rate Database
FG PLC DO (NDE) module	Fail to energize on demand	1.2E-07	7.4E-07	3.5E-07	39%	ESC Failure Rate Database
Line contactor (NDE)	Fails to close contacts on demand	0.0E+00	2.1E-07	9.0E-08	30%	Technis report T219
Safety timer relay	Contacts fail to open on demand	0.0E+00	1.5E-08	1.5E-06	99%	Technis report T219
Line contactor (NE)	Contacts fail to open on demand	0.0E+00	9.0E-08	2.1E-07	70%	Technis report T219

Table 16.5: Results.

Hazard	Target PFD	SIL	PTI hrs	PFD assessed	SIL from SFF	Overall SIL
Dam over-topping	5×10^{-3}	2	8760	4×10^{-3}	2	2
Water surge	2.3×10^{-3}	2	8760	4.6×10^{-4}	2	2

(d) Results and conclusions

The results of the assessment (Table 16.5) demonstrate that, based on the assumptions, the specified SIFs meet the hardware reliability and architectural requirements of the targets indentified by the LOPA.

Reliability block diagram, dam over-topping.

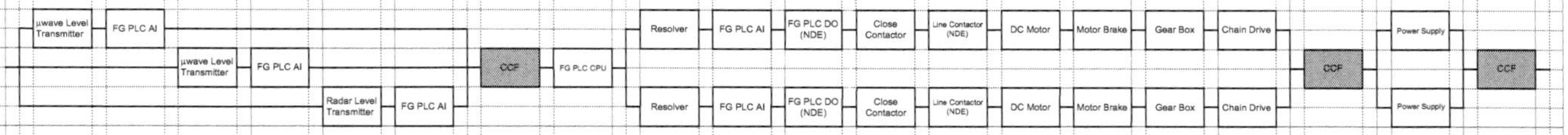

Dangerous failure mode:	Fails to detect high loch level	Fails to interpret high loch level	Fails to detect high loch level	Fails to interpret high loch level	Fails to detect high loch level	Fails to interpret high loch level		Fails to interpret high level or gate closed on demand	Erroneously detects gate in open positon	Erroneously detects gate in open positon	Fail to energise on demand	Fails to close contacts on demand	Fails to close contacts on demand	Fails to start on Demand	Fails on	Seizes	Breaks		Loss of power	
	µwave Level Transmitter	FG PLC AI	µwave Level Transmitter	FG PLC AI	Radar Level Transmitter	FG PLC AI	CCF	FG PLC CPU	Resolver	FG PLC AI	FG PLC DO (NDE)	Close Contactor	Line Contactor (NDE)	DC Motor	Motor Brake	Gear Box	Chain Drive	CCF	Power Supply	CCF
Configuration						2003	1001	1001		1002							1002	1001	1002	1001
CCF Contribution							5%											2%		10%
λDD (autotest)	9.94E-07	5.60E-07	9.94E-07	5.60E-07	1.13E-06	5.60E-07	8.42E-08	2.70E-07	1.35E-06	5.60E-07	1.23E-07	0.00E+00	0.00E+00	0.00E+00	0.00E+00	0.00E+00	0.00E+00	2.45E-09	5.50E-05	5.50E-06
λDD (branch)						3.67E-10	8.42E-08	2.70E-07									2.03E-06	2.45E-09	5.51E-05	5.50E-06
MDT						24	24	24									24	24	24	24
PFDdd						4.41E-09	2.02E-06	6.48E-06									2.38E-09	5.89E-09	1.75E-06	1.32E-04
λDU (proof test)	2.04E-07	2.11E-07	2.04E-07	2.11E-07	3.55E-07	2.11E-07	2.83E-08	3.00E-08	1.50E-07	2.11E-07	7.43E-07	2.10E-07	2.10E-07	3.25E-06	8.40E-08	2.70E-06	2.70E-06	1.98E-07	0.00E+00	0.00E+00
λDU (branch)						5.63E-09	2.83E-08	3.00E-08									1.03E-05	1.98E-07	0.00E+00	0.00E+00
Proof Test, Tp						8760	8760	8760									8760	8760	8760	8760
PFDdu						1.64E-05	1.24E-04	1.31E-04									2.69E-03	8.67E-04	0.00E+00	0.00E+00
PFDdd(sys)	1.42E-04																			
PFDdu(sys)	3.83E-03																			
PFD	3.97E-03																			
SIL (PFD)	SIL2																			
Type	B	B	B	B	B	B		B	B	B	A	A	A	A	A	A	A		A	
SFF	87%	82%	87%	82%	82%	82%		99%	95%	82%	39%	30%	30%	35%	30%	10%	10%		100%	
HFT	1	1	1	1	1	1		0	1	1	1	1	1	1	1	1	1		1	
Architectural SIL	2	2	2	2	2	2		2	3	2	2	2	2	2	2	2	2		4	
Allowed SIL (Arch)	**SIL2**																			

Reliability block diagram, water surge.

Dangerous failure mode:		Contacts fail to open on demand		Contacts fail to open on demand	
		Safety Timer Relay		Line Contactor	
Configuration		1oo1		1oo1	
CCF Contribution					
Qty		1		1	
λDD (autotest)		0.00E+00		0.00E+00	
DD (autotest) x Qty		0.00E+00		0.00E+00	
λDD (branch)				0.00E+00	
MDT				24	
λDDsys (autotest)				0.00E+00	
PFDdd				0.00E+00	
λDU (proof test)		1.50E-08		9.00E-08	
U (proof test) x Qty		1.50E-08		9.00E-08	
λDU (branch)				1.05E-07	
Proof Test, Tp				8760	
λDUsys (proof test)				1.05E-07	
PFDdu				4.60E-04	
λDD(sys) /hr		0.00E+00			
λDU(sys) /hr		1.05E-07			
λD(sys) /hr		**1.05E-07**			
PFDdd(sys)		0.00E+00			
PFDdu(sys)		4.60E-04			
PFD		**4.60E-04**			
SIL (PFD)		**SIL3**			
Type		A		A	
SFF		99%		70%	
HFT		0		0	
Architectural SIL		3		2	
Allowed SIL (Arch)		**SIL2**			

16.2 Spurious Opening of Either of Two Tidal Lock Gates Involving a Trapped Vessel

The scenario involves either one of a pair of lock gates moving despite no scheduled opening. This leads to a vessel becoming trapped and either sinking or causing harm to a person on board. A two-fatality scenario is perceived.

The following estimates of frequencies and propagations are credible:

Boat movements through the lock	12 p/day	Assume a half minute per passage
Boat situated such as to be trapped	17%	Based on an assumed 10 ft vessel in a 60 ft lock
Skipper fails to take avoiding action	10%	Judgement (noting 2 minutes closure time)
Entrapment causes damage to vessel	90%	Judged likely
Fatality ensues	50%	Judgement

The combination of the above factors, together with failures and incidents, is shown in Figure 16.1. The fault tree logic was analysed using the TECHNIS fault tree package TTREE, which is reproduced at the end of this chapter. The probability of the top event is $\mathbf{3.1 \times 10^{-5}}$.

Assuming a maximum tolerable risk of 10^{-5} pa for this involuntary public risk, the maximum tolerable failure rate for the mitigating effect of the Junction Gates is:

$$10^{-5}\ \text{pa}/3.1 \times 10^{-5} = \mathbf{3.2 \times 10^{-1}\ pa}.$$

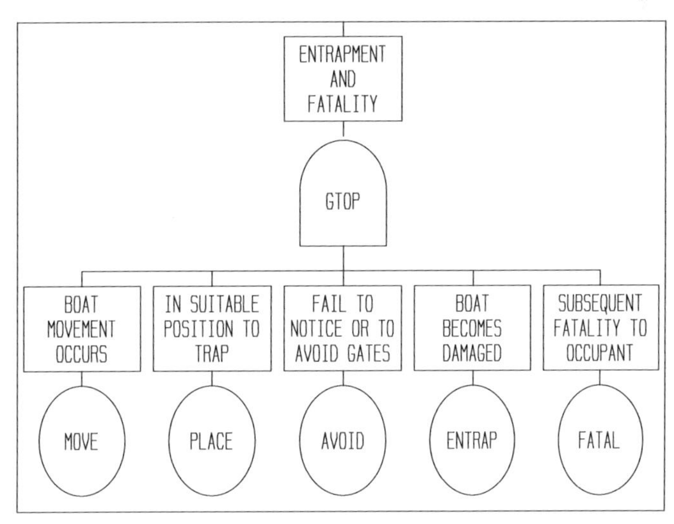

Figure 16.1: Fault tree.

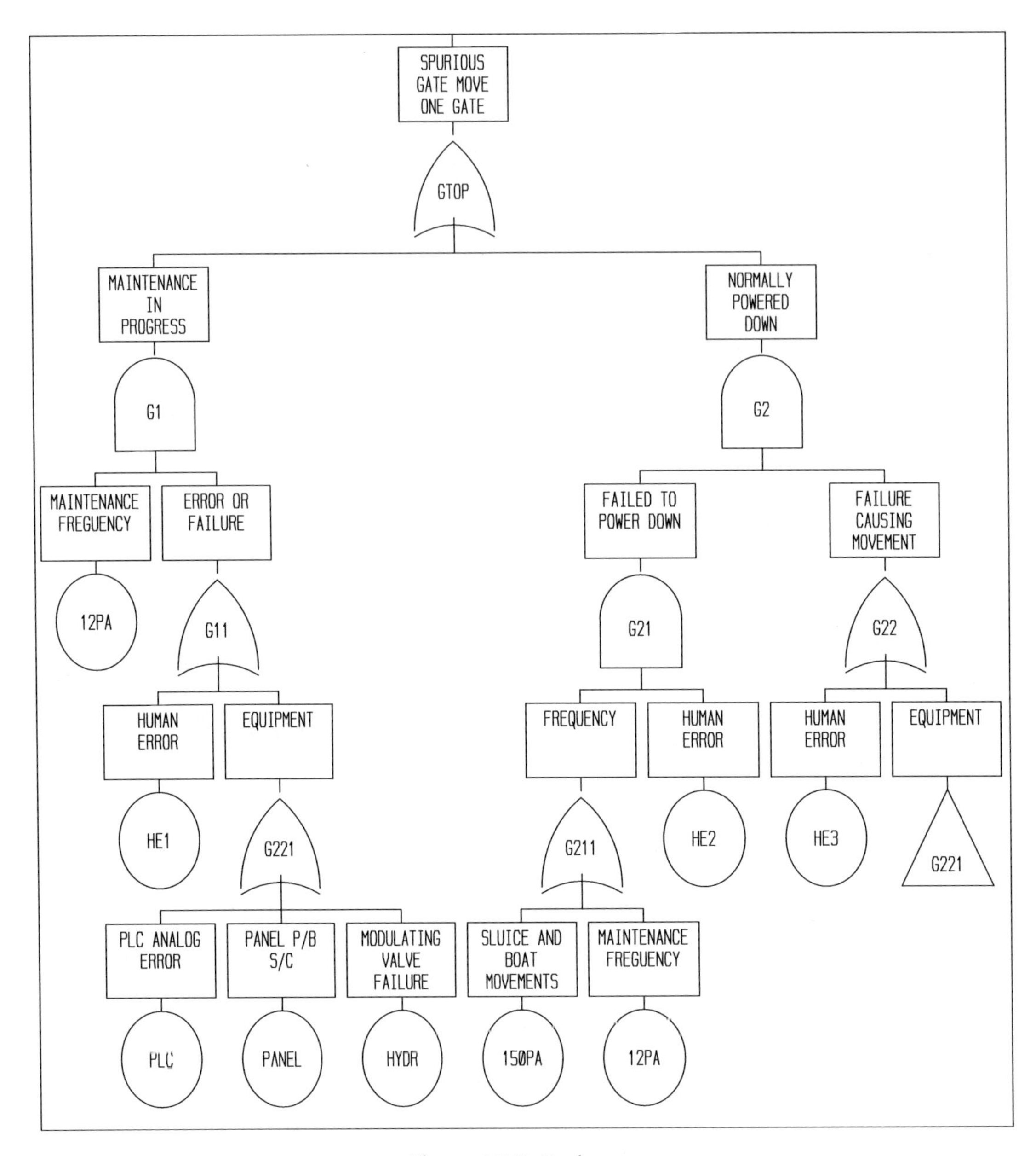

Figure 16.2: Fault tree.

The fault tree logic (Figure 16.2) was constructed as a result of studying the scenario. The frequency of the top event is $\mathbf{3.1 \times 10^{-1}}$ **pa** per gate, which meets the requirement.

The target (being greater than 10^{-1}) implies a target **<SIL 1**.

As can be seen from the fault tree output data shown at the end of this section, human error dominates the contributions to the top event (>95%).

We shall now address ALARP

A failure rate of 3.1×10^{-1} pa maps to a fatality risk of 10^{-5} pa $\times 3.1 \times 10^{-1}/3.2 \times 10^{-1}$

$= 9.7 \times 10^{-6}$ pa

Thus, assuming a "cost per life saved" criterion of £4,000,000, any proposal which might reduce the risk to the Broadly Acceptable limit of 10^{-6} pa might be tested as follows.

$$£4{,}000{,}000 = £\text{proposal}/(9.7 \times 10^{-6} - 10^{-6}) \times 30 \text{ yrs} \times 2 \text{ fatalities}$$

Thus any proposal costing less than **£2,300** should be considered. It is unlikely that any further risk reduction can be implemented within this sum; thus it might be argued that ALARP is satisfied.

However, it should be noted that:

- The predicted frequency is close to the target and reliability prediction is not a precise statistic
- The domination of human error suggests further investigation.

TTREE version 3.3 File name: T424D.TRO
Results of fault tree quantification for top event: GTOP
Top event frequency = 0.351E-04 per hour
0.307E+00 per year
Top event MDT = 0.105E+02 hours

Basic Event Reliability Data

Basic Event	Type	Failure Rate	Mean Downtime/ Test Interval		Constant Probability
12PA	I/E	.140E-02	12.0	(MDT)	
HE1	E				.200E-01
HE2	E				.200E-01
150PA	I/E	.170E-01	4.00	(MDT)	
HE3	E				.200E-01
PLC	I/E	.100E-05	336.	(PTI)	
PANEL	I/E	.100E-07	336.	(PTI)	
HYDR	I/E	.120E-05	336.	(PTI)	

Measure of cut set importance - ranked by frequency

Rank 1 Importance .785 MTBF hours.363E+05 MTBF years 4.15

Basic Event	Type	Failure Rate	Mean Downtime/ Test Interval		Constant Probability
12PA	I/E	.140E-02	12.0	(MDT)	
HE1	E				.200E-01

Rank 2 Importance .178 MTBF hours.160E+06 MTBF years 18.2

Basic Event	Type	Failure Rate	Mean Downtime/ Test Interval		Constant Probability
150PA	I/E	.170E-01	4.00	(MDT)	
HE2	E				.200E-01
HE3	E				.200E-01

Rank 3 Importance .147E-01 MTBF hours.194E+07 MTBF years 221.

Basic Event	Type	Failure Rate	Mean Downtime/ Test Interval		Constant Probability
12PA	I/E	.140E-02	12.0	(MDT)	
HE2	E				.200E-01
HE3	E				.200E-01

Rank 4 Importance .845E-02 MTBF hours.337E+07 MTBF years 385.

Basic Event	Type	Failure Rate	Mean Downtime/ Test Interval		Constant Probability
12PA	I/E	.140E-02	12.0	(MDT)	
HYDR	I/E	.120E-05	336.	(PTI)	

Rank 5 Importance .704E-02 MTBF hours.405E+07 MTBF years 462.

Basic Event	Type	Failure Rate	Mean Downtime/ Test Interval		Constant Probability
12PA	I/E	.140E-02	12.0	(MDT)	
PLC	I/E	.100E-05	336.	(PTI)	

APPENDIX 1

Functional Safety Management

Template Procedure

This procedure could be part of a company's Quality Management System (e.g. ISO 9001). It contains those additional practices (over and above ISO 9001) necessary to demonstrate Functional Safety Capability as would be audited by a reviewing body (see Chapter 7).

A large organization, with numerous activities and product types, might require more than one procedure, whereas a small company would probably find a single procedure satisfactory.

Again, the activities covered by a designer and manufacturer of instruments or systems will differ from those of a plant operator, which, in turn, will differ for a functional safety consultant/assessor.

This template has been successfully used by companies in the safety systems integration field and in consultancy firms. It consists of a top-level procedure and eight work practices to cover details of safety assessment (see Annex 1).

The terms used (e.g. Safety Authority, Safety Engineering Manager) are examples only, and will vary from organization; xxxs are used to designate references to in-house company procedures and documents.

This template should not be copied exactly as it reads but tailored to meet the company's way of operating.

Company Standard xxx Implementation of Functional Safety

Contents

Safety Critical Systems Handbook. DOI: 10.1016/B978-0-08-096781-3.10023-9

7. Functional Safety Specification
8. Life Cycle Activities

8.1 Integrity Targeting
8.2 Random Hardware Failures
8.3 ALARP
8.4 Architectures
8.5 Life-cycle activities
8.6 Functional Safety Capability

9. Implementation
10. Validation

Work Instruction xxx/001 – Random Hardware Failures & ALARP
Work Instruction xxx/002 – Integrity Targeting
Work Instruction xxx/003 – Life Cycle Activities
Work Instruction xxx/004 – Architectures (SFF)
Work Instruction xxx/005 – Rigour of Life Cycle Activities
Work Instruction xxx/006 – Functional Safety Competence
Work Instruction xxx/007 – Functional Safety Plan
Work Instruction xxx/008 – Functional Safety Specification

1 Purpose of document

This standard provides detail of those activities related to setting and achieving specific safety-integrity targets and involves the design, installation, maintenance and modification stages of the life-cycle. Where the activity in question is already catered for elsewhere in the XYZ Ltd quality management system, this document will provide the appropriate cross-reference.

The purpose of this procedure is to enable XYZ Ltd to provide in-house expertise in functional safety such as to meet the requirements of IEC 61508. Since IEC 61508 is not a prescriptive standard the issue is one of providing a risk based "safety argument" that is acceptable to one's regulator/auditor/HSE. A functional safety assessment consists of evidence showing that the areas of the standard have been adequately addressed and that the results are compatible with the current state of the art.

This requires a proactive risk-based approach rather than a slavish adherence to requirements.

2 Scope

The standard shall apply to all products and documentation designed, produced, installed or supported by XYZ Ltd except where contract requirements specifically call for an alternative.

In the case of simple designs, and modifications to existing plant, these activities may be carried using in house resources and skills. Larger projects may require the use of external resources.

Additional detail (to assist Project Safety Engineers or subcontractors) is supplied in Work Instructions/001 – /008.

The following diagram shows the relationship of relevant procedures:

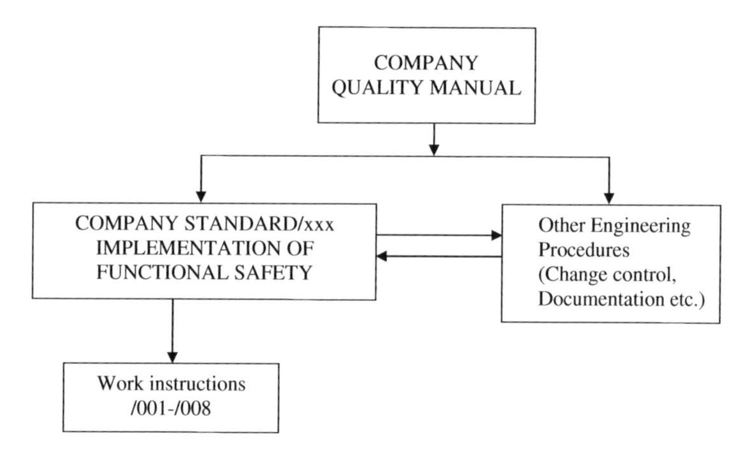

3 Functional safety policy

Paragraph x of the Quality Manual emphasizes that capability in respect of functional safety is a specific design capability within XYZ Ltd. Some contracts will relate to safety-related applications. Some developments will specifically target safety-integrity conformance as a design requirement.

If the project is deemed to be safety-related then the Project Manager shall appoint an independent Project Safety Assessor. However, a project may be declared sufficiently minor that formal hazard identification is not required and that the remainder of this procedure need not apply. That decision will only be undertaken or ratified by the Company Functional Safety Manager.

In the case of minor modifications this review process is satisfied by means of the impact analysis which shall be recorded on the change request.

4 Quality & safety plan

Every project shall involve a Quality & Safety Plan which is the responsibility of the Project Manager. It will indicate the safety-related activities, the deliverables (e.g. Safety-Integrity assessment report) and the competent persons to be used. The Project Manager will consult the competency register and will review the choice of personnel with the Safety Authority.

The tasks are summarized in Section 5 of this standard. Minimum SR items required in the Quality & Safety Plan are shown in WI/007.

See also Appendix 7 of this book

5 Competencies

The HR department will maintain a "safety-related competence register" containing profiles of those individuals eligible to carry out functional safety assessment and design tasks. Periodically the Managing Director and Functional Safety Manager will review the list.

The list will be updated from:

Individuals' attendance at relevant off-the-job courses
Records of SR experience from each project (on-the-job training) (Project Managers will provide this information to the Personnel Manager)
Details of new employees or contractors.

Sample entry in the competency register

See Chapter 2 Figure 2.5 of this book

Examples of specific jobs involving SR competencies include:

Functional Safety Manager

The FSM will provide the company's central expertise in functional safety. He/she will have substantial experience in functional safety assessment and will be thoroughly conversant with IEC 61508 and related standards.

Safety Authority:

This role requires the ability to bring to a project all the expertise necessary to define functional safety requirements and to carry out the assessments. He/she will communicate between disciplines on functional safety issues. The individual may not possess all the skills involved but is required to understand sufficient of the methodology to be able to manage the process of sub-contracting all or parts of the work. In other words, the competency to make valid judgments of the subcontracted work is of the essence. A minimum of one day's "off the job" training with a competent course-provider is required. He/she shall resolve conflicts with his/her other roles in the project by liaising with the Company Functional Safety Manager.

Functional Safety Auditor

Functional Safety Audits are carried out by a person other than the Safety Authority for a project. He/she will have received the XYZ Ltd training course on Functional Safety. He/she will have had experience of at least one Safety-Integrity Assessment.

Lead Project Engineer

A Lead Project Engineer shall have a basic understanding of the requirements of IEC 61508 such as might be obtained from a one-day appreciation course.

For each project, the Project Manager (assisted by the Safety Authority) shall consult the competence register to decide who will be allocated to each task. In the event that a particular competence(s) is not available then he will discuss the possible options involving training, recruitment or subcontracting the task with the Managing Director.

Each individual on the competency register will participate in an annual review (generally at the annual appraisal) with his/her next level of supervision competent to assess this feature of performance. He/she will also discuss his/her recent training and experience, training needs, aspirations for future SR work.

6 Review of requirements and responsibilities

6.1 Source of the requirement

There are two circumstances in which an integrity target will arise:

Arbitrary Requirement from a client with little or no justification/explanation
An integrity target based on earlier, or subcontractor, assessments. In the event of this being greater than SIL 1, derived from some risk graph technique, then XYZ Ltd should attempt to ratify the result by means of quantified risk targeting.

6.2 Contract or project review

Where a bid, or invitation to tender, explicitly indicates a SR requirement (e.g. reference to IEC 61508, use of the term safety-critical, etc.) then the Sales Engineer will consult a Safety Authority for advice.

All contracts (prior to acceptance by XYZ Ltd) will be examined to ascertain whether they involve safety-related requirements. These requirements may be stated directly by the client or may be implicit by reference to some standard. Clients may not always use appropriate terms to refer to safety-related applications or integrity requirements. Therefore, the assistance of the Safety Engineering Manager will be sought before a contract is declared not safety-related.

A project or contract may result in there being a specific integrity requirement placed on the design (e.g. SIL 2 of IEC 61508). Alternatively, XYZ Ltd may be required to advise on the appropriate integrity target in which case/002 will be used.

6.3 Assigning responsibilities

For each project or contract the Project Manager shall be responsible for ensuring (using the expertise of the Safety Authority) that the safety-integrity requirements are ascertained and implemented.

Each project will have a Safety Authority.

The Project Manager will ensure that the FS activities (for which he carries overall responsibility to ensure that they are carried out) called for in this standard (and related procedures) are included in the project Quality & Safety Plan and the life-cycle techniques and measures document. Specific allocation of individuals to tasks will be included in the Quality and Safety Plan. These shall include:

Design & implementation tasks
Functional safety assessment tasks
Functional safety audits.

The Project Manager will ensure that the tasks are allocated to individuals with appropriate competence. The choice of individual may be governed by the degree of independence required for an activity, as addressed in section 10 of this standard.

7 Functional safety specification

Every project shall involve a Functional Safety Specification. This is outlines in WI/008.

See also Chapters 2 and 4 of this book

8 Life-cycle activities

The IEC 61508 standard essentially addresses six areas:

Integrity targeting
Random hardware failures
ALARP
Architectures (safe failure fraction)
Life-cycle activities
Functional safety competence.

The life-cycle activities are summarized in this section They are implemented, by XYZ Ltd, by means of The Quality Management System (to ISO 9001 standard) by means of this standard and the associated Functional Safety Procedures (/001-008).

8.1 Integrity targeting

This is addressed in Chapter 2 of this book. The company choice of risks etc. will be described here

SIL 3 targets may sometimes be required but, for reasons of cost, additional levels of protection will be suggested. SIL 4 targets will always be avoided since they involve unrealistic requirements and can be better engineered by having additional levels of protection.

SIL targeting shall be carried out by using a quantified risk approach rather than any rule based risk graph methodology. In the event of an existing risk graph based assessment the

Company Functional Safety Manager shall advise that a risk based approach is necessary for functions indicated as greater than SIL 1 and will provide Company expertise in that area.

8.2 Random hardware failures

This involves assessing the design, by means of reliability analysis techniques, to determine whether the targets can be met. Techniques include fault tree and logic block diagram and FMEA analysis, redundancy modeling, assessments of common cause failure, human error modeling and the choice of appropriate component failure rate data. Reliability assessment may also be used to evaluate potential financial loss. The process is described in/001 (Random hardware failures).

8.3 ALARP (As Low As Reasonably Practicable)

This involves testing risk reduction proposals when the assessment of random hardware failures indicates that the target has been met but not by sufficient margin to reduce the risk below the broadly acceptable level.

It is necessary to decide whether the cost and time of any proposed risk reduction is, or is not, grossly disproportionate to the safety benefit gained. This requires that a cost per life (or non-injury) saved criterion is in place. The process is described in/001 (Random hardware failures).

8.4 "Architectures"

In the context of IEC 61508 the term "architectures" refers to the safe failure fraction parameter (or 2_H data route) for which there are SIL-related requirements. It involves establishing, for each piece of safety-related instrumentation, the fraction of failures which are neither unrevealed nor hazardous. The process is described in/004 (Architectures & safe failure fraction).

8.5 Life-cycle activities

In some cases existing safety assessments will have been based on only Integrity targeting, Random hardware failures, ALARP and Architectures (safe failure fraction). The Company Functional Safety Manager should advise that this represents only a part of the spectrum of functional safety assessment.

Where the Company Functional Safety Manager has made the decision to include an assessment of life-cycle rigor then the activities necessary to demonstrate conformance to a SIL target are summarized, in tabular form, in/005 – Life-cycle activities. Reference to the evidence which satisfies each requirement will be entered in the tables. Justifications for alternatives or for "not applicable" status will be entered in the same way.

Operations and Maintenance involve key activities which impact on the achievement of the functional safety targets. Specific items include:

Implementation of the correct proof test intervals as per the assessments
Recording all proof tests and demands on SIS elements.

8.6 Functional safety capability

8.6.1 Audit

The company has an ISO9001 QA audit capability and shall carry out at least one audit per annum of the implementation of this procedure.

8.6.2 Changes

Control of modifications is an important aspect and requires that all change request documents specifically identify changes as safety-related or NOT safety-related. The change request document will contain a "safety-related/not safety-related" option, a space to record the impact of the change. This judgement must be ratified by the Safety Authority.

8.6.3 Failures

Failure/defect/hazardous incident recording requires that each is identified as safety-related or NOT safety-related. This judgement must be ratified by the Safety Authority.

8.6.4 Placing requirements onto suppliers

Instrumentation and field devices

There is a need to place a requirement upon OEM suppliers defining the hazardous failure modes together with an integrity (e.g. SIL or SFF) requirement.

System integrators

Where a safety-related sub-system (e.g. F&GDS or ESD) is procured then a "Functional Safety Specification" shall be placed on the system-integrator (i.e. supplier). It will state the hazardous failure modes (e.g. Fail to respond to a pressure input) and provide integrity targets to be demonstrated by the supplier. The integrity targets should be expressed (for each hazardous failure mode) either as SIL levels or as specific failure rates or probability of failure on demand.

8.7 Functional safety assessment report

Throughout the life cycle there should be evidence of an ongoing assessment against the functional safety requirements. The assessment report should contain, as a minimum:

- Reason for the assessment
- Hazard and risk analysis if appropriate
- Definition of the safety related system and its failure modes

- Calculation of target SIL
- Reliability models and assumptions, for example down times and proof test intervals
- Failure data sources and reliability calculations
- Findings of the qualitative assessment of life-cycle activities
- A demonstration of rigor such as is described in Appendix 2 of this book
- Appropriate independence.

9 Implementation

During design, test and build, defects are recorded on "Defect Reports". During Site installation and operations they are recorded on "Incident Reports", which embrace a wider range of incident.

Problems elicited during design review will be recorded on form xxxx. Failures during test will be recorded as indicated in STD/xxx (factory) and PROC/xxx (Site).

All defect reports will be copied to the Functional Safety Manager, who will decide whether they are SR or not SR. He will positively indicate SR or not SR on each report. All SR reports will be copied to the Safety Authority, who will be responsible for following up and closing out remedial action.

All SR incident reports, defect reports and records of SR system demands will be copied to the XYZ Ltd Functional Safety Manager, who will maintain a register of failures/incidents. A 6-monthly summary (identifying trends where applicable) will be prepared and circulated to Project Managers and Technical Authorities and Safety Authorities.

10 Validation

Validation, which will be called for in the Quality & Safety Plan and is specified in section 7.4a) of this standard, will involve a Validation Plan. This plan will be prepared by the Safety Authority and will consist of a list of all the SR activities for the Project, as detailed in this standard and related procedures.

The Safety Authority will produce a Validation Report which will remain active until all remedial actions have been satisfied. The Safety Authority and Project Manager will eventually sign off the report, which will form part of the Project File.

Annex A

Notes on the Second-level Work Instructions 001-008

Work Instruction xxx/001 – Random Hardware Failures & ALARP

Will describe techniques to be used (see Chapters 5 and 6 and Appendix 4 of this book).

Work Instruction xxx/002 – Integrity Targeting

Will describe techniques and targets to be used (see Chapter 2 of this book).

Work Instruction xxx/003 – Life Cycle Activities

Will capture the tables from Chapters 2, 3,4 and 8 of this book.

Work Instruction xxx/004 – Architectures (SFF)

Will describe the rules from Chapters 3 and 8 of this book.

Work Instruction xxx/006 – Functional Safety Competence

Will provide the tasks and register format – see Chapter 2 of this book.

Work Instruction xxx/007 – Functional Safety Plan

See Appendix 7 of this book.

Work Instruction xxx/008 – Functional Safety Specification

See Chapters 3 and 4 of this book.

APPENDIX 2

Assessment Schedule

The following checklist assists in providing CONSISTENCY and RIGOR when carrying out an Integrity Assessment. The checklist can be used to ensure that each of the actions have been addressed. Furthermore it can be included, as an Appendix, in an assessment report with the Paragraph Numbers of the report referenced against each item. In this way a formal review of rigor can be included.

1 Defining the Assessment and the Safety System

1.1 Describe the reason for the assessment, for example safety case support, internal policy, contractual requirement for IEC 61508 Paragraph No.............................

1.2 Confirm the degree of independence called for and the competence of the assessor. This includes external consultants. Paragraph No.............................

1.3 Define the safety-related system. This may be a dedicated item of safety-related equipment (i.e. ESD) or a control equipment which contains safetyrelated functions. Paragraph No.............................

1.4 Define the various parts/modules of the system being studied and list the responsibilities for design and maintenance. For example, the PLC may be a proprietary item which has been applications-programmed by the supplier/user – in which case information will be needed from the supplier/user to complete the assessment. Paragraph No.............................

1.5 Describe the customer, and deliverables anticipated, for the assessment. For example "XYZ to receive draft and final reports". Paragraph No.............................

1.6 Provide a justification, for example that the SIL calculation yields a target of less than SIL 1, where it is claimed that an equipment is not safety-related. Paragraph No.............................

1.7 Establish that the development (and safety) life-cycle has been defined for the safety-related system. Paragraph.............................

1.8 Establish that the Quality Plan (or other document) defines all the necessary activities for realizing the requirements of IEC 61508 and that all the necessary design, validation, etc. documents are defined.

Safety Critical Systems Handbook. DOI: 10.1016/B978-0-08-096781-3.10024-0

2 Describing the Hazardous Failure Mode and Safety Targets

2.1 Establish the failure mode(s) which are addressed by the study, against which the safety-related system is deemed to be a level of protection (for example downstream overpressure for which ESD operates a slam-shut valve). Paragraph No..............................

2.2 Establish the risk criteria for the failure mode in question. Paragraph No..............................

2.3 Taking account of the maximum tolerable risk, calculate the SIL(s) for the safety-related system for the failure mode(s) in question. Indicate whether the SIL has been calculated from a risk target, for example Table 2.2 of Chapter 2 of this book, or derived from LOPA or risk matrix approaches. In the event of using risk graph methods, indicate the source and method of calibration of the method. Paragraph No ....

2.4 Check that the appropriate SIL table has been applied (High or Low demand). Paragraph No..............................

2.5 Review the target SIL(s) against the number of levels of protection and decide whether a lower SIL target, with more levels of protection, is a more realistic design option. Paragraph No..............................

2.6 Ensure that the design documentation, for example requirements specification, adequately identifies the use of the safety-related system for protection of the failure mode(s) defined. Paragraph No..............................

3 Assessing the Random Hardware Failure Integrity of the Proposed Safety-related System

3.1 Create a reliability model(s), for example fault tree, block diagram, event tree, for the safety-related system and for the failure mode(s) defined. Paragraph No..............................

3.2 Remember to address CCF in the above model(s). Refer to the literature for an appropriate model, for example BETAPLUS. Paragraph No..............................

3.3 Remember to quantify human error (where possible) in the above model(s). Paragraph No..............................

3.4 Remember to address both auto and manual diagnostic intervals and coverage in the above model(s). Paragraph No..............................

3.5 Select appropriate failure rate data for the model(s) and justify the use of sources. Paragraph No..............................

3.6 Quantify the model(s) and identify the relative contributions to failure of the modules/components within the SRS (safety-related system). Paragraph No.............................

3.7 Have any the SFF claims been justified or agued? Paragraph No.............................

4 Assessing the Qualitative Integrity of the Proposed Safety-related System

4.1 Check that the architectural constraints for the SIL in question have been considered and that the diagnostic coverage and safe failure fractions have been assessed. Paragraph No.............................

4.2 Review each paragraph of Chapters 3 and 4 of this book HAVING REGARD TO EACH FAILURE MODE being addressed. Remember that the qualitative feature applies to the safety-related system for a SPECIFIC failure mode. Thus, a design review involving features pertaining only to "spurious shutdown" would not be relevant where "failure to shutdown" is the issue. Paragraph No.............................

4.3 Document which items can be reviewed within the organization and which items require inputs from suppliers/subcontractors. Paragraph No.............................

4.4 Obtain responses from suppliers/subcontractors and follow up as necessary to obtain adequate VISIBILITY. Paragraph No.............................

4.5 Document the findings for each item above, and provide a full justification for items not satisfied but deemed to be admissible, for example non-use of Static Analysis at SIL 3 for a simple PLC. Paragraph No.............................

4.6 Has the use of software downloaded from a remote location, and any associated problems, been addressed? Paragraph No.........................

5 Reporting and Recommendations

5.1 Prepare a draft assessment report containing, as a minimum:

- Executive summary
- Reason for assessment
- Definition of the safety-related system and its failure modes
- Calculation of target SIL
- Reliability model
- Assumptions inherent in Reliability Model, for example down times and proof test intervals
- Failure data sources
- Reliability calculations
- Findings of the qualitative assessment

Report No.........................

5.2 If possible include recommendations in the report as, for example:

"An additional mechanical relief device will lower the SIL target by one, thus making the existing proposal acceptable".
"Separated, asynchronous PESs will reduce the CCF sufficiently to meet the target SIL".

Paragraph No..............................

5.3 Address the ALARP calculation where the assessed risk is greater than the broadly acceptable risk. Paragraph No..............................

5.4 Review the draft report with the client and make amendments as a result of errors, changes to assumptions, proposed design changes, etc.

Meeting (date) ....

6 Assessing Vendors

6.1 In respect of the items identified above requiring the assessment to interrogate subcontractors/suppliers, take account of other assessments that may have been carried out, for example IEC 61508 assessment or assessment against one of the documents in Chapters 8–10 of this book. Review the credibility and rigor of such assessments. Paragraph No..............................

6.2 In respect of the items identified above requiring the assessment to interrogate subcontractor/suppliers, ensure that each item is presented as formal evidence (document or test) and is not merely hearsay; for example "a code review was carried out". Paragraph No..............................

7 Addressing Capability and Competence

7.1 Has a functional safety capability (i.e. Management) review been conducted as per Chapter 2.3.5 of this book? Paragraph No..............................

7.2 Consider the competence requirements of designers, maintainers, operators and installers. Paragraph No..............................

7.3 Establish the competence of those carrying out this assessment. Paragraph..............................

APPENDIX 3

Betaplus CCF Model, Scoring Criteria

Checklist for Equipment Containing Programmable Electronics

A scoring methodology converts this checklist into an estimate of Beta. This is available as the Betaplus software package.

(1) Separation/segregation

Are all signal cables separated at all positions?
Are the programmable channels on separate printed circuit boards?
    OR are the programmable channels in separate racks
    OR in separate rooms or buildings?

(2) Diversity

Do the channels employ diverse technologies?
    1 electronic + 1 mechanical/pneumatic
    OR 1 electronic or CPU + 1 relay based
    OR 1 CPU + 1 electronic hardwired?
Were the diverse channels developed from separate requirements from separate people with no communication between them?
Were the two design specifications separately audited against known hazards by separate people and were separate test methods and maintenance applied by separate people?

(3) Complexity/design/application/maturity/experience

Does cross-connection between CPUs preclude the exchange of any information other than the diagnostics?
Is there >5 years experience of the equipment in the particular environment?
Is the equipment simple <5 PCBs per channel?
    OR <100 lines of code
    OR <5 ladder logic rungs
    OR <50 I/O and <5 safety functions?
Are I/O protected from over-voltage and over-current and rated >2:1?

Safety Critical Systems Handbook. DOI: 10.1016/B978-0-08-096781-3.10025-2

(4) Assessment/analysis and feedback of data

Has a combination of detailed FMEA, fault tree analysis and design review established potential CCFs in the electronics?
Is there documentary evidence that field failures are fully analysed with feedback to design?

(5) Procedures/human interface

Is there a written system of work on site to ensure that failures are investigated and checked in other channels? (including degraded items which have not yet failed)
Is maintenance of diverse/redundant channels staggered at such an interval as to ensure that any proof-tests and cross-checks operate satisfactorily between the maintenance?
Do written maintenance procedures ensure that redundant separations such as, for example, signal cables, are separated from each other and from power cables and should not be re-routed?
Are modifications forbidden without full design analysis of CCF?
Is diverse equipment maintained by different staff?

(6) Competence/training/safety culture

Have designers been trained to understand CCF?
Have installers been trained to understand CCF?
Have maintainers been trained to understand CCF?

(7) Environmental control

Is there limited personnel access?
Is there appropriate environmental control? (e.g. temperature, humidity)

(8) Environmental testing

Has full EMC immunity or equivalent mechanical testing been conducted on prototypes and production units (using recognized standards)?

Checklist and Scoring for Non-programmable Equipment

Only the first three categories have different questions as follows:

(1) Separation/segregation

Are the sensors or actuators physically separated and at least 1 metre apart?

If the sensor/actuator has some intermediate electronics or pneumatics, are the channels on separate PCBs and screened?

OR if the sensor/actuator has some intermediate electronics or pneumatics, are the channels indoors in separate racks or rooms?

(2) Diversity

Do the redundant units employ different technologies?

e.g. 1 electronic or programmable + 1 mechanical/pneumatic

OR 1 electronic, 1 relay based

OR 1 PE, 1 electronic hardwired?

Were separate test methods and maintenance applied by separate people?

(3) Complexity/design/application/maturity/experience

Does cross-connection preclude the exchange of any information other than the diagnostics?

Is there > 5 years experience of the equipment in the particular environment?

Is the equipment simple, e.g. non-programmable-type sensor or single actuator field device?

Are devices protected from over-voltage and over-current and rated >2:1 or mechanical equivalent?

(4) Assessment/analysis and feedback of data

As for Programmable Electronics (see above).

(5) Procedures/human interface

As for Programmable Electronics (see above).

(6) Competence/training/safety culture

As for Programmable Electronics (see above).

(7) Environmental control

As for Programmable Electronics (see above).

(8) Environmental testing

As for Programmable Electronics (see above).

The diagnostic interval is shown for each of the two (programmable and non-programmable) assessment lists. The **(C)** values have been chosen to cover the range 1–3 in order to construct a model which caters for the known range of BETA values.

For programmable electronics

Diagnostic coverage	Interval <1 min	Interval 1–5 mins	Interval 5–10 mins	Interval >10 mins
98%	3	2.5	2	1
90%	2.5	2	1.5	1
60%	2	1.5	1	1

For sensors and actuators

Diagnostic coverage	Interval <2 hrs	Interval 2 hrs – 2 days	Interval 2 days – 1 week	Interval >1 week
98%	3	2.5	2	1
90%	2.5	2	1.5	1
60%	2	1.5	1	1

The Betaplus model is available, as a software package, from the author

APPENDIX 4

Assessing Safe Failure Fraction and Diagnostic Coverage

In Chapter 3 safe failure fraction and random hardware failures were addressed and reference was made to FMEA.

1 Failure Mode and Effect Analysis

Figure A4.1 shows an extract from a failure mode and effect analysis (FMEA) covering a single failure mode (e.g. OUTPUT FAILS LOW).

Columns A and B identify each component.
Column C is the total part failure rate of the component.
Column D gives the failure mode of the component leading to the failure mode (e.g. FAIL LOW condition).
Column E = Column D × E shows the appropriate proportion of Column C (e.g. 20% for U8).
Column F shows the assessed probability of that failure being diagnosed. This would ideally be 100% or 0 but a compromise is sometimes made when the outcome is not totally certain.
Column H is a working column which multiplies the mode failure rate by the diagnostic coverage for each component.

Cells at the bottom of the spreadsheet in Figure A4.1 contain the algorithms to calculate diagnostic coverage (63%) and SFF (92%).

Diagnostic coverage is obtained from the sum of Column H divided by the sum of Column E.

SFF is obtained by taking the proportion of diagnosed hazardous failures (total of Column E × the diagnostic coverage) PLUS the failures deemed to be non-hazardous in this context (total of Column C minus total of Column E). This total is divided by the total failure rate (Column C) to obtain the SFF.

Typically this type of analysis requires 4 man-days of effort based on a day's meeting for a circuit engineer, a software engineer who understands the diagnostics and the safety assessor carrying out the "component by component" review. A further day allows the safety assessor to add failure rates and prepare the calculations and a report.

Safety Critical Systems Handbook. DOI: 10.1016/B978-0-08-096781-3.10026-4

XYZ MODULE							
A	B	C	D	E	F	G	H
COMP REF	DESCRIPTION	F.Rate	MODE 1	M1 F.rate	% Diag M1	NOTES	E*F
		pmh		pmh			
U6/7	2 x MOS LATCHES @ 0.01	0.02	20%	0.004	90		0.0036
U8	PROG LOGIC ARRAY	0.05	20%	0.01	0		0
U9-28	20 x SRAM @ .02	0.4	20%	0.08	90		0.072
U29-31	4 x FLASH MOS 4M @ .08	0.32	20%	0.064	50		0.032
TR23	npn lp	0.04	S/C	0.012	0		0
TOTAL F.Rate		0.83		0.17			0.1076
WEIGHTED %					63		
SFF =					92		

Figure A4.1: FMEA.

2 *Rigor of the Approach*

In order demonstrate the rigor of the FMEA exercise, Table A4.1 provides a template of items to be addressed. It can thus be used, in the FMEA report, to indicate where each item can be evidenced.

Table A4.1: Rigor of the FMEA.

A definition of the equipment's intended safety function and perceived failure mode	Section? of this
Summary of failure data used	Section? of this
General and specific assumptions	Section? of this
Spreadsheet (or FARADIP output) showing, for each failure mode of the equipment, the component failure rates and modes (for each block identified in the reliability/fault model) the data source used (with any justifications if necessary)	Section? of this
Where the FMEA involves more than one block, the reliability or fault tree models showing the architecture, common cause failures (if redundant), equations used, calculations, MTTR and proof test interval, etc.	Section? of this
Justification for any diagnostic coverage claimed for each component (if over 60%). This may involve a separate textual section describing the hardware/software/watchdog arrangements	Section? of this
Where applicable, the predicted effect of temperature variation on the failure data used, (e.g. elevated temperature approaching the maximum junction temperature).	Section? of this
Where applicable, factoring of the failure rate data, used where components (such as power transistors, electrolytic capacitors) have been used above 70% of their rated load	Section? of this

Table A4.1 ***(Continued)***

Where applicable, factoring of the failure rate data, to allow for the effect of high vibration	Section? of this
Identification of any life limited components, together with maintenance/replacement requirements (e.g. batteries, electrolytic capacitors, electro-mechanical components, etc.).	Section? of this
Documented evidence of a theoretical circuit design review (showing scope, findings, reviewer independence, etc.)	Section? of this
Circuit design information: Schematics, including block diagram if multi-board Parts list Functional description including on-board diagnostics (if any)	Section? of this
Safety requirements specification and/or brief product specification (e.g. datasheet) including environmental and application information	Section? of this

APPENDIX 5

Answers to Examples

Answer to Exercise 1 (Chapter 2.1.1d)

Propagation to fatality is 1:2 times 1:5 = 0.1.
Maximum tolerable failure rate leading to single fatality is 10^{-5} pa / 10^{-1} = 10^{-4} pa; however the actual process failure rate is 0.05 pa = 5×10^{-2} pa.
Thus the protection system should have a target probability of failure on demand (pfd) no worse than:

10^{-4} pa/5×10^{-2} pa = 2×10^{-3}.

The target is dimensionless and is thus a PFD. the Low Demand column in Table 1.1 is therefore indicated.
Thus the requirement is SIL 2.

Answer to Exercise 2 (Chapter 2.1.1d)

Answer 2.1

Since there are 10 sources of risk (at the same place) the maximum tolerable fatality rate (per risk) is $10^{-5}/10 = 10^{-6}$ pa.
Target toxic spill rate is 10^{-6} pa/ 10^{-1} = 10^{-5} pa.
However, the actual spill rate is 1/50 pa = 2×10^{-2} pa.
Thus the protection system should have a target probabilty of failure on demand no worse than:

10^{-5} pa/2×10^{-2} pa = 5×10^{-4}.

The target is dimensionless and is thus a PFD. The Low Demand column in Table 1.1 is therefore indicated.
Thus the requirement is SIL 3.

Answer 2.2

The additional protection reduces the propagation to fatality to 1:30 so the calculation becomes: target spill rate is 10^{-6} pa/ 3.3×10^{-2} pa = 3×10^{-5} pa; however. spill rate is 1/50 pa = 2×10^{-2} pa.
Thus the protection system should have a target probabilty of failure on demand no worse than:

3×10^{-5} pa/2×10^{-2} pa = 1.5×10^{-3}.

Thus the requirement now becomes SIL 2 (low demand).

Safety Critical Systems Handbook. DOI: 10.1016/B978-0-08-096781-3.10027-6

Answer to Exercise 3 (Chapter 2.1.1d)

Target maximum tolerable risk = 10^{-5}pa.
Propagation of incident to fatality = 1/200 = 5×10^{-3}.
Thus target maximum tolerable failure rate =10^{-5}pa / 5×10^{-3} = 2×10^{-3} pa:
note: 2×10^{-3} pa = 2.3×10^{-7} per hour.
The requirement is expressed as a rate, thus the High Demand column of Table 1.1 is indicated at SIL 2.

Answer to Exercise 4 (Chapter 2.2)

For the expense to just meet the cost per life saved criterion then:
£2,000,000 = £proposal / $(8 \times 10^{-6} - 2 \times 10^{-6}) \times 3 \times 25$ = £900.
Thus an expenditure of £900 would be justified if the stated risk reduction could be obtained for this outlay. Expenditure greatly in excess of this could be argued to be disproportionate to the benefits.

Answer to Exercises (Chapter 11)

11.2 Protection system

The target Unavailability for this "add-on" safety system is therefore 10^{-5}pa/2.5×10^{-3}pa = $\mathbf{4 \times 10^{-3}}$, which indicates **SIL 2**.

11.4 Reliability block diagram

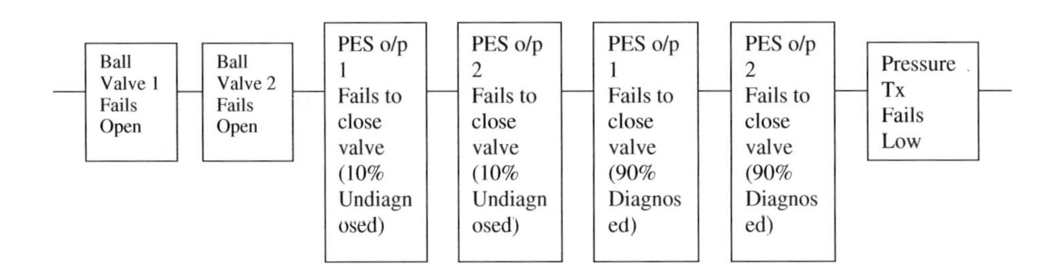

11.6 Quantifying the model

(a) Ball valve SS1 fails open.
Unavailability = λ MDT = $0.8 \times 10^{-6} \times 4000$
$= 3.2 \times 10^{-3}$

(b) Ball valve SS2 fails open.
Unavailability = λ MDT = $0.8 \times 10^{-6} \times 4000$
$= 3.2 \times 10^{-3}$

(c) PES output 1 fails to close valve (Undiagnosed Failure).
Unavailability = 10% λ MDT = $0.025 \times 10^{-6} \times 4000$
$= 1 \times 10^{-4}$

(d) PES output 2 fails to close valve (Undiagnosed Failure).
Unavailability = 10 % λ MDT = $0.025 \times 10^{-6} \times 4000$
$= 1 \times 10^{-4}$

(e) PES output 1 fails to close valve (Diagnosed Failure).
Unavailability = 90% λ MDT = $0.225 \times 10^{-6} \times 4$
$= 9 \times 10^{-7}$

(f) PES output 2 fails to close valve (Diagnosed Failure).
Unavailability = 90% λ MDT = $0.225 \times 10^{-6} \times 4$
$= 9 \times 10^{-7}$

(g) Pressure transmitter fails low
Unavailability = λ MDT = $0.5 \times 10^{-6} \times 4000$
$= 2 \times 10^{-3}$

The predicted Unavailability is obtained from the sum of the unavailabilities in (a) to (g)

= 8.6×10^{-3}. (Note: the target was 4×10^{-3}.)

This is higher than the unavailability target. The argument as to the fact that this is still within the SIL 2 target was discussed in Chapter 2. We chose to calculate an unavailability target and thus it is NOT met.

74% from items (a) and (b), the valves.

23% from item (g), the pressure transmitter.

Negligible from items (c)–(f), the PES.

11.7 Revised diagrams

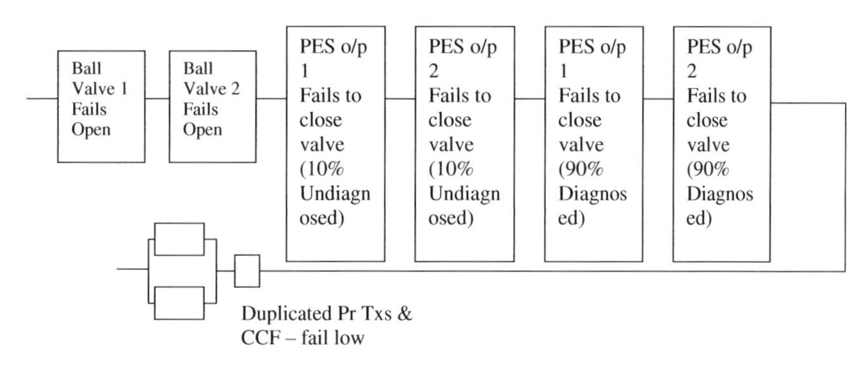

Reliability block diagram.

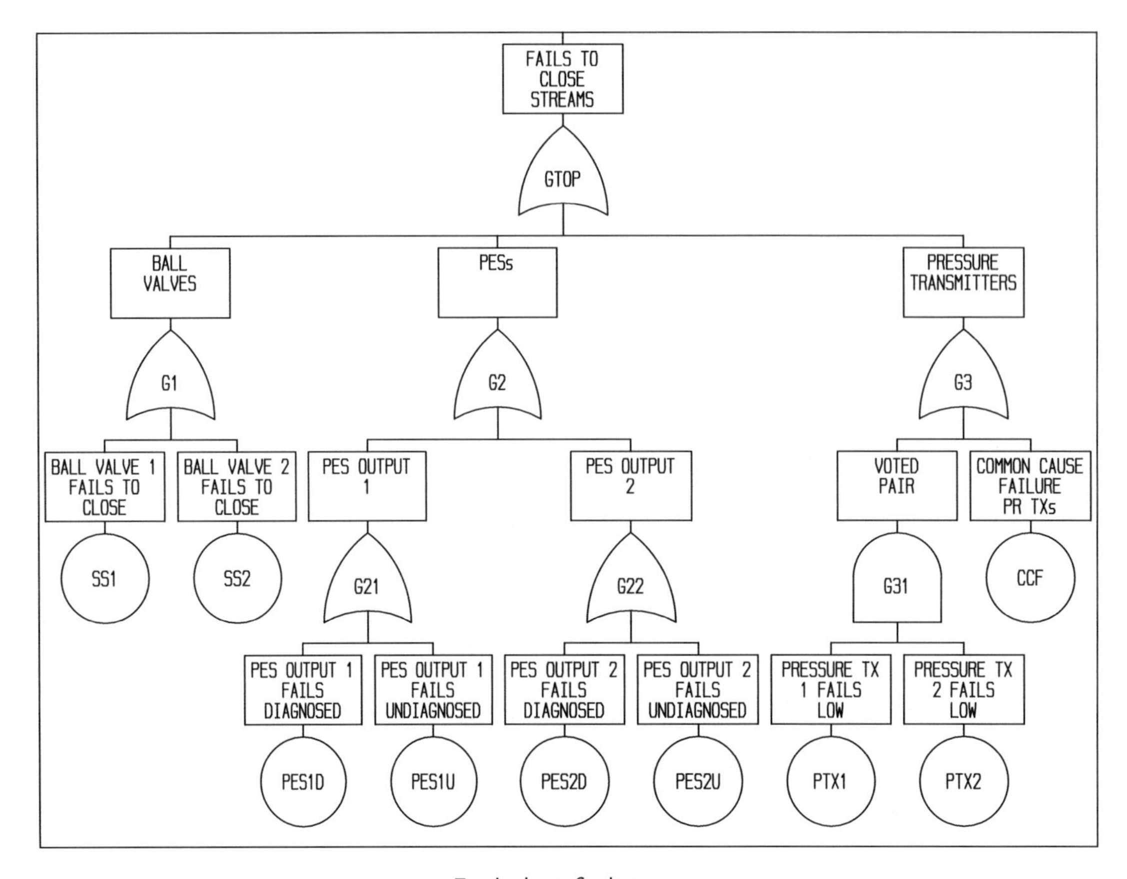

Equivalent fault tree.

11.9 Quantifying the revised model

Changed figures are shown in bold.

(a) Ball valve SS1 fails open.

Unavailability = λ MDT = $0.8 \times 10^{-6} \times \mathbf{2000}$

$= \mathbf{1.6 \times 10^{-3}}$

(b) Ball valve SS2 fails open.

Unavailability = λ MDT = $0.8 \times 10^{-6} \times \mathbf{2000}$

$= \mathbf{1.6 \times 10^{-3}}$

(c) PES output 1 fails to close valve (Undiagnosed Failure).

Unavailability = 10% λ MDT = $0.025 \times 10^{-6} \times \mathbf{2000}$

$= \mathbf{5 \times 10^{-5}}$

(d) PES output 2 fails to close valve (Undiagnosed Failure).

Unavailability = 10 % λ MDT = $0.025 \times 10^{-6} \times$ x **2000**

$= \mathbf{5 \times 10^{-5}}$

(e) PES output 1 fails to close valve (Diagnosed Failure).
  Unavailability = 90% λ MDT = $0.225 \times 10^{-6} \times 4$
  $= 9 \times 10^{-7}$

(f) PES output 2 fails to close valve (Diagnosed Failure).
  Unavailability = 90% λ MDT = $0.225 \times 10^{-6} \times 4$
  $= 9 \times 10^{-7}$

(g) Voted pair of pressure transmitters.
  Unavailability = $\lambda^2 T^2/3 = [0.5 \times 10^{-6}]^2 \times 4000^2/3$
  $= 1.3 \times 10^{-6}$

(h) Common cause failure of pressure transmitters.
  Unavailability = 9% λ MDT = $0.09 \times 0.05 \times 10^{-6} \times 2000$
  $= 9 \times 10^{-5}$

The predicted Unavailability is obtained from the sum of the unavailabilities in (a) to (h) = **3.3×10^{-3}**, which meets the target.

11.10 ALARP

Assume that further improvements, involving CCF and a further reduction in proof test interval, could be achieved for a total cost of £1,000. Assume, also, that this results in an improvement in unavailability, of the safety-related system, from **3.3×10^{-3}** to the PFD associated with the Broadly Acceptable limit of **4×10^{-4}**. It is necessary to consider, applying the ALARP principle, whether this improvement should be implemented.

If the target unavailability of **4×10^{-3}** represents a maximum tolerable risk of **10^{-5} pa** then it follows that **3.3×10^{-3}** represents a risk of $10^{-5} \times 3.3/4 =$ **8.3×10^{-6} pa**. If **10^{-6}** pa is taken as the boundary of the negligible risk then the proposal remains within the tolerable range and thus subject to ALARP.

Assuming a two-fatality scenario, the cost per life saved over a 40 year life of the equipment (without cost discounting) is calculated as follows:

3.3×10^{-3} represents a risk of 8.3×10^{-6}
4×10^{-4} represents a risk of 10^{-6}
Cost per life saved = £1,000 / (40×2 lives $\times$ [8.3-1] $\times 10^{-6}$)
= £1,700,000

On this basis, if the cost per life saved criterion were £1,000,000, then justification for the further improvement would be considered marginal as the benefit is just below (but close to) the criteria. On the other hand it would be justified if the criterion were £2,000,000.

11.11 Architectural constraints

(a) PES
The safe failure fraction for the PESs is given by 90% diagnosis of 5% of the failures, which cause the failure mode in question, PLUS the 95% which are "fail safe".
Thus (90% × 5%) + 95% = 99.5%.
Consulting the tables in Chapter 3 then:
If the simplex PES is regarded as Type B then SIL 2 can be considered if this design has >90% safe failure fraction.

(b) Pressure transmitters
The safe failure fraction for the transmitters is given by the 75% which are "fail safe".
If they are regarded as Type A then SIL 2 can be considered since they are voted and require less than 60% safe failure fraction.
Incidentally, in the original proposal, the simplex pressure transmitter would not have met the architectural constraints.

(c) Ball valves
The safe failure fraction for the valves is given by the 90% which are "fail safe".
If they are are regarded as Type A then SIL 2 can be considered since they require more than 60% safe failure fraction.

Comments on Example (Chapter 12)

The following are a few of the criticisms which could be made of the Chapter 12 report.

12.2 Integrity requirements

In Chapter 11 the number of separate risks to an individual was taken into account. As a result the 10^{-4} pa target was amended to 10^{-5} pa. This may or may not be the case here but the point should be addressed.

12.4.1 ALARP

It was stated that nothing could be achieved for £672. It may well be possible to achieve significant improvement by reducing proof test intervals for a modest expenditure.

12.5 Failure rate data

It is not clear how the common cause failure proportion has been chosen. This should be addressed.

Other items

(a) There is no mention of the relationship of the person who carried out the assessment to the provider. Independence of the assessment needs to be explained.
(b) Safe failure fraction was not addressed.
(c) Although the life-cycle activities were referred to, the underlying function safety capability of the system provider was not called for.

APPENDIX 6

References

ANSI/ISA-84.00.01 (2004) Functional Safety, Instrumented Systems for the Process Sector.

CWA 15902-1 (2009) Lifting and Load Bearing Equipment for Stages and other Production Areas within the Entertainment Industry.

EEMUA Guidelines – Publication No 160, 1989, Safety related instrument systems for the process industry (including programmable electronic systems).

EN ISO 14121 Principles of Risk Assessment – Machinery.

EN 15998 Earth Moving Machinery – MCS using Electronics

EN 62061 - Functional Safety of E/E/PES – Machinery.

EN ISO 13849 - Safety Related Parts of Control Systems – Machinery.

ISO/DIS 26262 Road Vehicles - Functional Safety

ISO/DIS 25119 Tractors and Machinery for Agriculture

EN 474 Earth Moving Machinery – Safety

EN 50126 Draft European Standard: Railway applications - The Specification and Demonstration of Dependability, Reliability, Maintainability and Safety (RAMS).

EN 50128 - Software for railway control and protection systems.

EN 50129 - Hardware for railway control and protection systems.

EN 60204-1 Safety of machinery – electrical equipment of machines.

EN 61800-5-2:2007 Adjustable speed electrical power drive systems.

EN 954-1 Safety of machinery in safety-related parts of control systems.

Guide to the Application of IEC 61511 to safety instrumented systems in the UK process industries.

Gulland W G, Repairable redundant systems and the Markov Fallacy. Journal of Safety and Reliability Society Vol 22 No 2 Summer 2002.

HSE, 1992, Tolerability of risk for nuclear power stations, UK Health and Safety Executive, ISBN 0118863681. *Often referred to as TOR.*

HSE, 2001, Reducing risks, protecting people. *Often referred to as R2P2.*

HSE, 1995, Out of Control: Control systems: why things went wrong and how they could have been prevented. HSE Books, ISBN 0 7176 0847 6.

HSE 190, 1999, Preparing safety reports: Control of Major Accident Regulations. Appendix 4 addresses ALARP.

HSE, 2000, Regulating higher hazards: Exploring the issues.

HSE Publication, 1989, Guidance on the use of programmable electronic systems in safety-related applications.

HSE (2007) Managing competence for safety-related systems.

IEC 60601 Medical Electrical Equipment, General requirements for basic safety and essential performance

IEC Standard 61508, 2010 Functional safety: safety related systems - 7 parts.

IEC Standard 61713, 2000 Software dependability through the software life-cycle processes – Application guide.

IEC Standard 62061 Safety of machinery – functional safety of electronic and programmable electronic control systems for machinery.

IEC Standard 61511: Functional safety – safety instrumented systems for the process industry sector.

IEC Draft International Standard 61513: Nuclear Power Plants – Instrumentation and control for systems important to safety – general requirements for systems.

IEC Publication 61131, Programmable Controllers, 8 Parts, (Part 3 is programming languages).

IET Publication, 1992, Guidelines for the documentation of software in industrial computer systems, 2nd edition, ISBN 08634104664.

Safety Critical Systems Handbook. DOI: 10.1016/B978-0-08-096781-3.10028-8

IET/BCS, 1999, Competency guidelines for safety-related system practitioners, ISBN 085296787X.
ISO/DIS 25119, Tractors and Machinery for Agriculture ISO13849.
Institution of Gas Engineers & Managers publication IGEM/SR/15, Programmable equipment in safety related applications, Edition 5, 2010.
Instrument Society of America, S84.01, 1996, "Application of Safety Instrumented Systems for the Process Industries", ISBN 1556175906.
MISRA (Motor Industry Software Reliability Assoc), 1994, Development guidelines for vehicle based software, ISBN 0952415607.
MISRA (Motor Industry Software Reliability Assoc) Coding Standard
MISRA (Motor Industry Software Reliability Assoc) 2007, Guidelines for safety analysis of vehicle based software, ISBN 97809524156-5-7
Norwegian Oil Ind Assoc, OLF-070, Recommended guidelines for the application of IEC 61508 in the Petroleum activities on the Norwegian Continental Shelf.
RSSB Engineering Safety Managemet (The Yellow Book), Issue 4.0, ISBN 9780955143526.
RTCA DO-178B/(EUROCAE ED-12B), 1992, Software considerations in airborne systems and equipment certification.
RTCA/DO-254 Design Assurance Guidance for Airborne Electronic Hardware.
Simpson K G L, Reliability assessments of repairable systems – is Markov modelling correct? Journal of Safety and Reliability Society Vol 22 No 2 Summer 2002.
Smith D J, 2011, Reliability, Maintainability and Risk, 8th edition (Elsevier), ISBN 9780080969022.
Smith D J, FARADIP.THREE, Version 6.5, 2010, User's manual, Reliability software package, ISBN 0 9516562 3 6.
Smith D J, BETAPLUS Version 3.0, 1997, User's manual, Common cause failure software package, ISBN 09516562 5 2.
Smith D J, 2000, Developments in the use of failure rate data and reliability prediction methods for hardware, ISBN 09516562 6 0.
Storey N, 1996, Safety critical computer systems, Addison Wesley, ISBN 0201427877.
Technis Guidelines Q124, 2004, Demonstration of Product/System Compliance with IEC 61508.
UKAEA, 1995, Human reliability assessors guide (SRDA-R11), June 1995, Thomson House, Risley, Cheshire WA3 6AT, 0853564205.
UK MOD Interim Defence Standard 00-55: The procurement of safety critical software in defence equipment.
UK MOD Interim Defence Standard 00-56 (Issue 4.0 – 2007): Safety Management Requirements for Defence Systems.
UK MOD Interim Standard 00-58: A guideline for HAZOP studies on systems which include programmable electronic systems.
UK MOD Interim Defence Standard 00-54: Requirements for safety-related electronic hardware in defence equipment.
UL (Underwriters Laboratories Inc, USA), 1998, Software in programmable components, ISBN 0 76290321X.

APPENDIX 7

Quality and Safety Plan

Typical items for inclusion are:

1 Responsibilities (by name and must be listed in the company competency register)

Project manager
Functional safety authority for the project
Functional safety audit

2 Life-cycle Details

Overall life-cycle, e.g. design, control of sub-contract software, test, installation and commissioning.
Software life-cycle (see Chapter 4) including tools and compilers and their version numbers.

3 Hazard Analysis and Risk Assessment

Allocation of targets to sub-systems.
Description of failure modes (e.g. spurious valve movement, spurious release, loss of heating, overpressure etc)
SIL targets (e.g. SIL 1 for functions A, B & C and SIL 2 for functions D & E).

4 Items/deliverables to be Called for and Described in Outline

Document Hierarchy

e.g. requirements specification, hardware specs & drawings, software spec, code listings, review plan and results, test plan and results, validation plan and report, relevant standards such as for coding or for hardware design.

List of Hardware Modules

Including the configuration of hardware (e.g. voted channels and redundant items). Details of their interconnection and human interfaces.

Safety Critical Systems Handbook. DOI: 10.1016/B978-0-08-096781-3.10029-X

List of Software Items

Media, listings.

Safety Manual

Hardware and/or software manual.

Review Plan

e.g. design reviews of functional spec and of code listings and test results and validation report.

Test Plan

e.g. List of module tests, functional test, acceptance tests, environmental tests.

Validation Plan/report

Could be in the form of a matrix of rows containing the numbered requirements from the functional or safety spec and columns for each of the reviews, tests, assessments etc.

5 Descriptions of

The boundary of the safety-related system (e.g. input and output signals relating to the safety functions).

APPENDIX 8

Some Terms and Jargon of IEC 61508

The 7 "Parts" of IEC 61508 are described as **"normative"**, which means they are the Standard proper and contain the requirements which should be met. Some of the annexes, however, are described as **"informative"** in that they are not requirements but guidance which can be used when implementing the normative parts. It should be noted that the majority of Parts 5, 6 and 7 of the Standard are informative annexes.

A few other terms are worth a specific word or so here:

Functional safety is the title of this book and of IEC 61508. It is used to refer to the reliability (known as integrity in the safety world) of safety-related equipment. In other words it refers to the probability of it functioning correctly, hence the word "functional".

E/E/PE (electrical/electronic/programmable electronic systems) refers to any system containing one or more of those elements. This is taken to include any input sensors, actuators, power supplies and communications highways. Providing that one part of the safety-related system contains one or more of these elements the Standard is said to apply to the whole.

ELEMENT: one or more components providing a safety function or part thereof.

EUC (equipment under control) refers to the items of equipment which the safety-related system being studied actually controls. It may well be, however, that the EUC is itself safety-related and this will depend upon the SIL calculations described in Chapter 2.

FSCA – see FSM.

FSM: functional safety management (previously referred to as functional safety capability assessment (FSCA).

HR and R are used (in IEC 61508) to refer to "Highly Recommended" and "Recommended". This is a long-winded way of saying that HR implies activities or techniques which are deemed necessary at a particular SIL and for which a reasoned case would be needed for not employing them. R implies activities or techniques which are deemed to be "good practice".

NR is used to mean Not Recommended, meaning that the technique is not considered appropriate at that SIL.

Safety Critical Systems Handbook. DOI: 10.1016/B978-0-08-096781-3.10030-6

SOUP: software of unknown pedigree.

Verification and validation: verification (as opposed to validation) refers to the process of checking that each step in the life-cycle meets earlier requirements. Validation (as opposed to verification) refers to the process of checking that the final system meets the original requirements.

Type A components (hardware or software): implies that they are well understood in terms of their failure modes and that field failure data is available. See Chapter 3.

Type B components (hardware or software): implies that any one of the Type A conditions is not met. See Chapter 3.

Should/shall/must: in standards work the term "must" usually implies a legal requirement and has not been used in this book. The term "shall" usually implies strict compliance and the term "should" implies a recommendation. We have not attempted to differentiate between those alternatives and have used "should" throughout this book.

Software packages

FARADIP.THREE (£450)

Described in Chapter 5, a unique failure rate and failure mode data bank, based on over 40 published data sources together with Technis's own reliability data collection. FARADIP has been available for 25 years and is now widely used as a data reference. It provides failure rate DATA RANGES for a nested hierarchy of items covering electrical, electronic, mechanical, pneumatic, instrumentation and protective devices. Failure mode percentages are also provided.

TTREE (£775)

Used in Chapters 12–16, a low-cost fault tree package which nevertheless offers the majority of functions and array sizes normally required in reliability analysis. TTREE is highly user-friendly and, unlike more complicated products, can be assimilated in less than an hour. Graphical outputs for use in word processing packages.

Betaplus (£125)

Described in Chapter 5 and in Appendix 3, Betaplus has been developed and calibrated as a new-generation common cause failure partial β model. Unlike previous models, it takes account of proof-test intervals and involves positive scoring of CCF-related features rather than a subjective "range score". It has been calibrated against 25 field data results, obtained by Technis, and has the facility for further development and calibration by the user.

Available from:

TECHNIS, 26 Orchard Drive, Tonbridge, Kent TN10 4LG

Tel: 01732 352532
Fax: 01732 360018
Technis.djs@virgin.net

Reduced prices for combined packages or for software purchased with training courses
(Prices at time of press)

Index